𝕻𝖆𝖑𝖆𝖔𝖓𝖙𝖔𝖌𝖗𝖆𝖕𝖍𝖎𝖈𝖆𝖑 𝕾𝖔𝖈𝖎𝖊𝖙𝖞, 1919.

A MONOGRAPH

OF THE

FOSSIL INSECTS

OF THE

BRITISH COAL MEASURES.

BY

HERBERT BOLTON, M.Sc., F.R.S.E., F.G.S.,

DIRECTOR OF THE BRISTOL MUSEUM.

PART I.

PAGES 1—80, PLATES I—IV.

LONDON:

PRINTED FOR THE PALÆONTOGRAPHICAL SOCIETY

JULY, 1921.

PRINTED BY ADLARD AND SON AND WEST NEWMAN LTD. LONDON AND DORKING

THE FOSSIL INSECTS OF THE BRITISH COAL MEASURES.

INTRODUCTION

THE first recorded Palæozoic insect of any country appears to have been a British specimen, *Lithosialis brongniarti* (Mantell), which was discovered in the Coal Measures of Coalbrookdale in the early part of last century. It was sent by Mantell to Brongniart as a leaf impression. Brongniart in turn submitted the fossil to Mons. Audouin who (1833, Audouin, 'Ann. Soc. Ent. France, ii, Bull., p. 7) described it as "d'un insecte inconnu" and allied to the Hemerobiidæ, *Semblis*, and especially to *Corydalis* and *Mantis*. The specimen was afterwards figured and named by Mantell (1854, 'Medals of Creation,' vol. ii, p. 575, fig. 2).

According to Parkinson, however, Lhuyd first recognised fossil insects in the British Coal Measures. Parkinson ('Organic Remains, vol. iii, p. 258, 1804—1811) states that Lhuyd in a postscript to a letter to Dr. Richardson wrote as follows: "Scripsi olim suspicari me Araneorum quorundam icones una cum lithophytis, in schisto carbonario observasse, hoc jam ulteriore experientia edoctus aperte assero. Alias icones habeo, quæ ad Scarabæorum genus quam proxime accedunt. In posterum ergo non tantum Lithophyta, sed et quædam insecta in hoc lapide investigare conabimur." ('Lithophylacii,' p. 113.)

[" I have formerly written that I believed I had observed certain impressions of spiders identical with Lithophytes in carbonaceous shales, this I now, taught by later experience, openly assert. I have other impressions which approach nearest to the family of beetles. For the future, therefore, we will endeavour to investigate not only Lithophyta, but also certain insects in these shales."]

Parkinson reprints four figures given by Lhuyd in his 'Iconograph,' tab. 4. Two of these figures show eight legs and must therefore represent the remains of Arachnids. None of the figures show wing-structure.

Interest in the occurrence of fossil insects was stimulated in 1837 by the publication of Dean Buckland's 'Bridgewater Treatise' on Geology, in which he described and figured two fossils found at Coalbrookdale by Mr. Anstice (1837, Buckland, 'Geology and Mineralogy,' 2nd ed., vol. ii, p. 76). He determined both specimens to be the remains of coleopterous insects—a determination since corrected by H. Woodward (1871, 'Geol. Mag.,' vol. viii, p. 386, pl. xi), by Scudder, and finally by Pocock who referred them to the Arachnida ("Terrestrial Carboniferous Arachnida," 'Mon. Pal. Soc.,' 1911, pp. 39, 77).

1

Attention was afterwards diverted from the Coal Measures by the remarkable discoveries of insect-remains made by Brodie in the Purbeck and Liassic rocks, and by similar discoveries on the continent.

Mr E. W. Binney (1867, 'Proc. Lit. and Phil. Soc. Manchester,' vol. vi, p. 59) exhibited a specimen which "bore some resemblance to the pupa state of a coleopterous insect" and had been found in the Cinderford Dyke Pit at Bradley, near Huddersfield.

A second specimen exhibited by Binney at the same time was referred to *Xylobius sigillariae*, Dawson. Binney added: "We must expect great additions to be made to the Carboniferous fauna, as doubtless the rich and luxurious vegetation of that remote period would afford food and shelter for numerous insects."

Binney's notes on the Huddersfield specimens caused the Rev. P. B. Brodie to record (1867, 'Geol. Mag.' vol. iv, pp. 285—286) that he had in his collection "a wing of a gigantic Neuropterous insect in ironstone from the Derbyshire Coal Measures."

The same year Kirkby (1867, 'Geol. Mag.' vol. iv, pp. 388—390) reported the finding of clearly defined insect-remains in the Durham Coal Measures. One example consisted of "portions of the fore-wing or tegmina of an orthopterous insect nearly allied to *Blatta* or Cockroach," and the other of an orthopterous insect, apparently the abortive wing of a specimen related to the Phasmidae.' Kirkby's first specimen is the small but very fine wing described here under the name of *Phylomylacris mantidioides* (Sternberg). The second specimen is not determinable as an insect-fragment, and may prove to be a fossil fruit, referable to one of the higher plants of the Coal Measures.

From 1867 onwards the finding of fossil insects in the British Coal Measures occurred at long intervals until in 1908, the date of publication of Handlirsch's 'Fossilen Insekten,' the following had been recorded.

Phylomylacris mantidioides (Goldenberg)	(*Blattoidea*) *peachi* (Woodward)
olim, "allied to *Blatta*," Kirkby	olim, '*Etoblattina peachi*, Woodward
olim "*Blattina mantidioides* "	*Leptoblattina exilis* Woodward
Goldenberg	*Lithomylacris kirkbyi*, Woodward
Lithosialis brongniarti (Mantell)	*Scaamylacris deareusis* (Woodward)
olim, "*Gryllacris brongniarti*" Mantell	olim "*Etoblattina dearcusis*, Woodward
Lithomantis carbonarius, Woodward	*Pseudetoagnea cambrensis* (Allen)
Archaeoptilus ingens Scudder	olim, *Fouquea cambrensis* Allen
Brodia priscotincta Scudder	*Breyeria woodwardiana* (Handlirsch)
Eubleptus aegidea, Scudder	olim affinity with *Lithomantis carbonarius*,'
Aphthoroblattina johnsoni (Woodward)	Stobbs
olim, *Etoblattina johnsoni*, Woodward	olim, *Stobbsia woodwardiana*, Handlirsch

The numerous discoveries of insect-remains in the Coal Measures of Commentry (Allier), France, and the remarkable series made known by Handlirsch from the continent generally, and from the United States, overshadowed the limited British series, which seemed almost trivial by comparison. The present monograph shows, however, that the fossil insect-remains of the British Coal Measures are far more abundant than was supposed, and that they are by no means unimportant. About seventy specimens are known of about sixty species, and they represent five of the great groups of fossil insects.

Palæodictyoptera are the dominant forms, and are closely followed by Blattoids, several of which are referable to genera occurring in the French and Belgian coalfields. The British examples of *Soomylacris* are represented near Lens and Laévin by *Soomylacris hermensis*, Pr , while *Phylomylacris mantidioides* has its counterparts in *Phylomylacris nodosa*, Pr , and *Phylomylacris lafittei*, Pr , from Lens. The great Protodonata of Commentry are represented in the Bristol coalfield by *Boltonites radstockensis*.

The generic identity of French and British Coal Measure insects implies that they formed part of a general and wide-spread fauna, a view which is strengthened by the fact that while *Soomylacris deanensis* and *S. stocki* occur in the Forest of Dean coalfield, to the west, *S. barri* occurs in the Kent coalfield to the east, and only separated by the Straits of Dover from the Coal Measures of Northern France, in which Pruvost finds other species of the same genus.

Pruvost has also shown that in the Coal Measures of Lens and Laévin there is present a well-defined horizon of *Anthracomya phillipsii*, in which that species passes through the same developmental changes as in the Kent coalfield.

It is extremely likely that the Kent coalfield will later yield numerous insect-remains closely allied to those of France, and that systematic search will amplify the list of forms already known from all the British coalfields.

The British Palæodictyoptera on the whole, are more varied than the French, few forms showing the primitive condition of *Stenodictya*, while certain examples, such as *Megmoptera tuberculata*, *Palæomantis macroptera*, and the three genera of Lithomantids, are highly specialised. A similar degree of specialisation is seen in the British examples of the Protorthoptera, while the Blattoids, by their numerous genera and species, indicate that the group had a long history and a wide geographical range in the British coal period.

The fossil insects already found in the British Coal Measures form probably but a small fraction of those which remain to be discovered when attention is more fully directed to them. The insect-fauna, however is not usually associated with the general fauna in the Coal Measures, but occurs in beds of lighter coloured rock than the ordinary carbonaceous shales, and with abundant ironstone nodules or in the case of the Blattoids, in association with masses of drifted vegetation in the black shales, where the nervation of the wings so closely simulates the pinnules of *Neuropteris* as to be mistaken for the latter and cast aside.

While insect-remains are usually regarded as wholly restricted to the Westphalian and Stephanian stages of the Coal Measures the fauna with which they are most often associated in this country may indicate a greater age as it is known to occur as low down as the Calciferous Series of Scotland. Any statement, therefore, of the range in time of British Palæozoic insects based on the present known forms may have to be set aside by later discoveries.

The fauna with which fossil insects are usually associated in Great Britain is one in which arachnids and certain of the more primitive arthropods are the dominant forms. Arachnids are known to occur in the Calciferous Sandstone Series of Scotland at Redhall, near Slateford, Edinburgh, and in the Cement-stone Group of the Lower Carboniferous at Langholme, Dumfriesshire (1911, Pocock, ' Mon Pal Soc,' p 18), and elsewhere the genus *Archæoctonus* being represented by *A glober* and *A tuberculatus*, and the genus *Cyclophthalmus* by *C euglyptus* at Redhall, Blan Point, near Dysart, and Cramon near Edinburgh.

No insect-remains are known from any of these horizons, but if the faunal association seen in the Coal Measures is a trustworthy guide they may be looked for with some prospect of success.

The faunal association existing in the ' Seapstone Bed ' of the Lower Coal Measures at Carre Heys, Colne, Lancashire (1905, Bolton, ' Geol Mag ' [5], vol ii), is so similar in character to the typical insect-fauna elsewhere, that it is likely that insects lived in the Lower Coal Measure period in Lancashire.

The faunal association at Carre Heys is as follows and may be compared with the faunal association in which insects have been found to occur in other coalfields

ARTHROPODA
  Prycephalus cooperi Huxley
  Anthrapalemon serratus Woodw
          woodwardi Etheridge
          tiaquairi Peach
  Prestwichia rotundata, Woodw
  Archatarbus subovalis Woodw
  Euphoberia browni, Woodw
  Xylobius mondifer Woodw

PISCES
  Hybodopsis wardi, Barkas
  Acanthodes wardi Egerton
  Platombtleys athena Traq

AMPHIBIA
  Loxomma alldi (A S Woodw)
  Microsaurian remains

The oldest known fossil insect in the British Carboniferous appears to be a fragmentary wing (*Genentomum subaculum*), described by the author from shales at a depth of 637 feet below the Bedminster Great Vein in the Bristol Coalfield, and therefore at a considerable depth below the Pennant Grit

Pseudofouquea cambrensis (Allen) was obtained from the top of the Four foot Seam in the Lower Coal Measures at the Llanbradach Colliery near Cardiff, while the shales over the No 2 Rhondda Seam have yielded a wing-fragment of *Bottonella tenuitegminata* (Bolton), and the shales over the Graigola Seam have yielded the wings of two Blattoids, *Hemiuglaris conteça* and *Orthomylacris lanceolata*

The No. 2 Rhondda Seam and the Graigola or Six-foot Vein of Swansea both occur in the Pennant Grits, the former near the base of the series and the latter at 200 yards below the Swansea Four-foot Seam which forms the base of the Middle Coal Measures in South Wales.

In Monmouthshire, insect-remains occur in shales over the Mynddslwyn Vein, a seam at the base of the Upper Coal Measures.

The Durham and South Lancashire Coalfields have yielded insect-remains in measures near the top of the Middle Series, while those recorded from the Derbyshire Coalfield are on a still lower horizon in the Middle Series. Few fossil insects are known from the Upper Coal Measures.

HABITS AND MODE OF OCCURRENCE OF FOSSIL INSECTS

The bodies, and more particularly the wings, of insects, have been entombed in various deposits under conditions difficult to determine. Whatever the conditions they must have been closely related to the habits of life. The older writers claimed that wind-dispersal and water-carriage were the chief agents. Buckland, for example ('Anniv. Address to the Geol. Soc., 1842), supposed "that multitudes of insects have been occasionally drifted by tempests to the sea." Mantell ('Wonders of Geology,' 7th ed, 1857) pointed out that Westwood had drawn attention to the fact that "the streams brought down innumerable insects at certain periods, perhaps those of heavy rain."

Alfred Russel Wallace ('The Geographical Distribution of Animals, 1876) and Heilprin ('The Distribution of Animals,' 1887) alike drew attention to the widespread occurrence of living insects far out at sea, in some instances still flying strongly. Members of the British Association on their voyage to Australia in 1914 verified these statements by the capture of locusts as their vessel proceeded down the Red Sea and into the Indian Ocean. More than a score of locusts were captured on the vessel by which the writer travelled, and many more must have been driven down into the water by the fringe of a simoon into which the vessel entered beyond Aden.

Many insects are destroyed yearly by falling into streams and rivers after the deposition of their eggs in the water, and by becoming entangled in the surface film.

The occurrence of whole, or almost whole, insects is more likely to furnish surer proof of the conditions under which life was passed than is the occurrence of wings only, because the bodies, being more compact and much heavier than the wings, are less likely to have drifted to great distances. Sometimes, as we shall see later when we consider special cases, such as the Coal Measures of Commentry, France, or the remarkable faunal associations of certain of the British fossil insects, valuable information is supplied by the deposits, or by the nature of the associated forms of life.

The great group of the Palæodictyoptera and certain of the Protorthoptera and Protodonata had large wings, and were powerful fliers. We should therefore expect to find their remains widely dispersed in deposits of varied nature. This seems to be the case. Compact heavy-bodied insects like the Blattoids would have a more limited range, and their bodies after death could not be carried to great distances. Larval forms would in most cases be included in the deposits in the immediate neighbourhood of the area in which they lived.

M. Henri Fayol in his description of the Coal Measures of Commentry, France shows that these deposits were laid down in narrow land-locked lakes of a trough-like form lying in depressions of older schistose and crystalline rocks. The tranquil waters received only the finest mud in suspension, and the resultant mudstones have yielded a large insect-fauna, in which Blattoids are most numerous. The bodies of the insects are preserved in many cases. Certain of the insects were strong fliers, and their occurrence with the bodies intact indicates that they in all probability haunted the vicinity of the lakes and flew over them. When strongly-flying insects like *Bollonites radstockensis* or *Lithosialis brongniarti* died upon the land, the wings, because of their membranous and chitinous nature, would persist after the destruction of the softer body, and be swept off into streams after heavy rains or flooding of the land-surface, their great superficial area combined with their lightness making flotation easy.

The transference of insect-wings from the land into water would be accompanied by the drifting of plant-material, and the two would be buried together in the deposit then forming. The wing of *Bollonites* from Radstock was found with plant-remains in deposits of this sort, and may be taken as a proof supported as it is by other examples of Protodonate wings, that these insects lived over the land and died upon it.

The Palæodictyoptera, with their wings capable only of an up-and-down movement in one plane at right-angles to the body, and, when in a position of rest, disposed straight outwards, are not likely to have frequented the ground except in the open. These insects, like most of the Palæozoic forms, were all of large size, as contrasted with living types. Pruvost assumes that the characters of the wing unfitted these insects for a forest life, and that they must have been restricted to flight in the open neighbourhood of swamp pools. I do not wholly agree with this assumption, for the branches and leaves of the Coal Measure plants do not seem to have had so great a density and interlacing of foliage as seriously to impede the flight of powerful winged insects. There seems no reason why these insects should not have lived among the brakes of Lepidodendroid and Calamitean trees, and after death fallen or been blown into adjacent waters. The fact that isolated wings are often found in perfect condition and without any signs of wear and fear such as the wings of aged insects show to-day raises the question whether in some cases the wings were not shed, as in certain species of recent ants the shorn insect continuing its life as a ground-feeder.

The Protorthoptera were judging from the structure of their mouth-parts, somewhat general feeders or carnivorous, and the presence of strong walking legs suggests that they spent much of their life on the ground, possibly along the margins of swamps, where food would be especially abundant. They had nevertheless powerful wings, and some members, such as the Œdischidæ had legs adapted for leaping. Orthoptera, represented chiefly by Blattoid forms, were all fitted for flight by means of their large membranous hind-wings, and equally well fitted by powerful walking legs for life on the ground. In repose the hind-wings were hidden under the modified fore-wings.

I have elsewhere given my reason for a belief that the Blattoids were not wholly phytophagous, but in all probability carnivorous also ('Quart. Journ. Geol. Soc.,' vol. lxvii. p. 153, 1911).

Blattoids may also have entered the water in search of food, for the hind flying wings would be securely protected by the tegmina, whose broad muscular bases of attachment were sufficiently powerful to compress them down on the back and prevent water entering beneath just as in the case of the living water-beetle, *Hydrophilus piceus*. The chitinous surface of the body and of the tegmina would not hinder progress in water for their surfaces are no rougher than those of the modern *Dytiscus*, nor would the insect on emerging bring with it so heavy a film of water as to clog its movements. An objection may be found in the presence of stout bristle-like hairs on the legs seen on such forms as *Necymylacris lenulei* (Bolton) (1917, 'Mem. and Proc. Lit. and Phil. Soc. Manchester,' vol. lxi, p. 15), which might conceivably cause air-bubbles to cling in such profusion as to prevent the insect being able to submerge. The presence of fine hairs on the swimming legs of *Hydrophilus* and *Dytiscus* does not hinder the immersion of these insects in water, so that this is not a valid objection. If no hindrance to immersion was caused by the bristle like hairs, they may have been useful in assisting the act of swimming.

I think the probabilities are in favour of the Blattoids being at least semi-aquatic as well as land insects.

CONDITIONS OF LARVAL INSECT LIFE

The conditions under which larval life was passed are even more conjectural. The Protodonata may be regarded as insects whose larvæ must have been aquatic, like the aquatic larvæ of the Odonata now living, but Tillyard (1917, 'The Biology of Dragonflies,' Camb. Univ. Press, p. 306) conjectures that since adult Protodonata are found at Commentry without the occurrence of larval forms, the latter

may have dwelt in damp earth rather than in water and that "the formation of
the larval tracheal system undoubtedly proves that this at one time was the case
It may well have been so in Carboniferous times."

Tillyard's views are well worth quoting in full especially as they support in
some measure those of Pruvost "We may picture to ourselves the giant insects
of Commentry as inhabiting the shores of a large, shallow, nearly stagnant
lake   In the muddy ooze around its borders grew forests of the Giant Mare's
Tail while further back on the sandy slopes the graceful Cycads and other extra-
ordinary plants formed a more diversified medley There, amidst rotting vegetation,
these insects lived and bred   In such almost amphibious conditions it may well be
that the larvæ of Protephemeroidea and Protodonata first began that series of
adaptive changes which finally led them to adopt a purely aquatic mode of life"

The larvæ of *Brodia* and of other forms whose wings I describe under the
name of '*Pteronepionites*' must have lived under conditions fitted for their gradual
metamorphosis   The body was long well segmented, and bore rudimentary
wings which were carried well up over the thorax in an erect or semi-erect position
Though rudimentary, the wings possess features which may have determined to a
large extent the mode of life   They are attached by broad strong bases to the
thorax, and are very muscular, as shown by the stout ridges proceeding from the
point of attachment into the wings, and the anterior margins are also thickened
The bodies with their lateral expansions of the terga are very suggestive of those
of the Diplopoda, and like them would offer no serious obstacle to progression
through rank and rotting vegetation   That these larval insects would also
penetrate soft muds, if necessary, in search of food is possible, since the
soft-bodied caterpillars of the Hawk-moths of to-day are able to enter hard soil
before pupation takes place

The stout wing bases and the strengthened margins of the wings would
prevent damage to these structures as the larvæ crawled about, or sought to bury
themselves in the soil or muds   They were essentially adapted for a ground
habit   Whether they were capable of an aquatic or semi-aquatic habit can only
be settled by a knowledge of the mode of respiration

Lubbock, Gegenbaur and others have adduced strong reasons in favour of an
aquatic origin of the insects and in the Carboniferous types we should naturally
expect that the original habits had not had time to undergo any great modification
Larval wings of the '*Pteronepionites*' type must have been living structures in
which metabolism was active, and very unlike the dried membranous sac-like
expansions of the adult insect   The growth of the larval wings was continued
throughout metamorphosis, and during this period their delicate nature broad
expanse and the thinness of the integument may have enabled them to assist in
the respiratory function

The researches of Comstock and Needham show that larval wings of recent insects receive a plentiful supply of tracheal branches at an early stage, and it is evident that these are much in excess of any aeration the wings are likely to require. The tracheal development seems to be a persistence from an earlier more active condition, when the larval wings may have played a part in assisting the respiration. These considerations, and the presence of spiracle-like structures in the interstitial neuration of the adult wings of many Palæodictyoptera lend support to the inference that the wings functioned as organs of respiration. These spiracle-like structures are usually oval or rounded in outline and thickened. In some instances they show a series of raised lines radiating from the thickened edge into the surrounding areas as if they had been muscular strands and capable of expansion and contraction.

Are these structures the atrophied remains of spiracles once functional, and fitting the larva for a more or less aquatic existence? During a recent visit to this country Dr Tillyard has suggested to me that they are rudiments of sensory organs which may have been scent-glands. Scent-glands are known to occur on the wings of many insects, as, for example, the Green-veined White (*Pieris napæ*) the Small White (*Pieris rapæ*) and others, and their appearance is certainly a strong argument in favour of the view. Scent-glands are, however, in all probability but specialised developments of previously existing structures, and it is possible that the glandular-like organs to which I give the name of "pseudo-spiracles," and Handlirsch the name of "pterostigmata," are an earlier development connected with the "tracheoles" of Comstock, or that primary tracheolation to which Tillyard has given the name of "archyodictyon." Tillyard does not accept the view that they had ever any connection with respiration.

The almost total absence of structures which can be accepted as functional gills in these fossil insect-larvæ may be accounted for by the perishable nature of such organs. Before dismissing the question of the respiratory function in its relation to the conditions under which larval life was passed, it is desirable to draw attention to the larval Blattoid *Leptoblattina eallis* Woodw. In this insect the abdominal segments have the dorsal hinder margin extended into broad lamellar expansions so filmy in texture that they may have served as organs of aeration. The lamellar expansions were longer in life than they now are, the hinder borders showing an irregular torn edge. Their extreme thinness would permit of a ready osmotic-like action, especially in damp vegetation, or in an aquatic or semi-aquatic habitat. Scudder, Handlirsch, Lameere and others are all agreed in the belief that the Blattoids frequented decaying vegetation in or near water, and under these conditions the presence of organs of aeration similar in character to the abdominal expansions of *Leptoblattina eallis* would be of the greatest value and offer no difficulties to the habit of life. No similar structures are known in any other larval Blattoid, so that the view cannot be pressed.

2

We shall not be far wrong in assuming that the larvæ of some of the Coal Measure insects were wholly aquatic, others semi-aquatic—that the adult Blattoids were indifferently aquatic or terrestrial, the adults of the non-Blattoid types spent most of their life in the vicinity of the swamp-pools in which their larval stages were passed, and to which they might need to return to lay their eggs.

Such a view seems to accord with the known facts, and will explain the special character of the fauna of such deposits as those of Coseley, in Staffordshire, and the brick-clays of Sparth Bottoms Rochdale Lancashire. These are evidently true lagoon or swamp-pool deposits as contrasted with the ordinary shales and binds of the Coal Measures.

FOOD OF COAL MEASURE INSECTS

The nature of the food of Coal Measure insects has been much discussed, as it is so closely associated with habits. Handlirsch considers that the Gymnosperms and Pteridosperms of the Coal Measure forests were not likely to have been frequented by insects in search of food, as these plants do not prove attractive to living insects. Pruvost, on the other hand (1919 - 1920, "La Faune Continentale du Terrain Houiller du Nord de la France," Mém. Carte Géol. France, pp. 266—267), considers that many members of the Coal Measure flora possessed in their spores, or in the case of the higher plants, in their cones, a plentiful food supply for insects, and he finds in the association of a *Phylloblatta* at Lens with the *Potonica* of *Lanopteris* some support for his conclusions.

The contemporaneous rapid development of plants and insects is also quoted by Pruvost in support of his views.

Several writers have argued that the powerful wings and consequent powers of rapid flight of many of the insects are more in accordance with a predatory and carnivorous habit than with a purely frugivorous or herbivorous one, and this belief has led Lameere to write as follows (1917, 'Bull. Soc. Zool. France,' vol. xlii, pp. 36—37): "Over the lake of Commentry flew magnificent Ephemeroptera and splendid Odonatoptera the carnivorous larvæ of which were aquatic—doubtless the Odonatoptera, when fully grown devoured the Ephemeroptera, of which the most fully developed types, the Megasecoptera [or] which have left no descendants, must have made great slaughter among the smaller insects.

"On the ground in the forests, swarmed innumerable Blattoids which frequented the detritus, and which had as enemies the ferocious and agile Orthoptera the varied counterparts of our Mantidæ. These latter must have attacked equally the large vegetarian Orthoptera, the counterparts of the Phasmæ, which probably climbed on trees, and the bulky Protohemiptera, which sucked the sap. Some of the Orthoptera jumped and there were some which by their appearance recall our Acridians; but all these beings were mute.

"Small amphibians, and numerous arachnidians came to limit the swarm of this articulate world in a country without birds or mammals"

Pruvost does not accept the view of a primitive aquatic origin of insects, but affirms his belief in a terrestrial origin, and thinks that even if an aquatic habit be proved for certain of the larvæ of the Coal Measure insects, the habit must have been secondary, and derived from an earlier land ancestry (*op cit*, p 268)

Scudder has observed in the case of the Blattidæ that the venation of the tegmina very closely resembles the surface-features of the *Neuropteris* pinnule—so strongly in fact as to suggest mimetism

Pruvost rightly urges that a mimetism is of little value unless the mimetic insect frequents the plant mimicked At the same time, it can be urged that the stout compact bodies of the fossil Blattoids and their powerful walking legs were equally admirably fitted for progression among rank and decaying vegetation, and that in these conditions the Blattoids were quite as likely to have been omnivorous, while finding some degree of protection among the *Neuropteris* pinnules lying on the ground

The writer has previously commented on the association of the wings of Blattoids with the leaves of *Cordaites* (1911, 'Quart Journ Geol Soc ' vol lxvii, pp 161—165), and has made the following comment "While Carboniferous Blattoids may have been wholly phytophagous, it is interesting to note that the leaves of *Cordaites* (in the present case) are impressed with shallow pits, which show faint traces of a spiral I have in many previous instances found that such pits owed their origin to attached shells of *Spirorbis pusillus* Whether these leaves were partially submerged in water during life is an open question ; but in all cases the plant-tissues of the pittings are depressed, and are accurate impressions of *Spirorbis* If the Carboniferous Blattoids were not wholly vegetable feeders, the occurrence of *Spirorbis pusillus* upon the *Cordaites* may supply a reason for their frequent association '

CLASSIFICATION

The classification of fossil insects has presented great difficulties, both to the palæontologist, and to the systematist of living forms Palæozoic insects show to the systematist a series of forms not strictly referable to any modern grouping, but presenting certain generalised details of structure which link two or more now widely separated groups, besides other features not met with in living forms

The palæontologist finds that he has not to deal with early and primitive types, followed by a regular series showing a developmental progression, but with an apparent sudden incursion of large series of highly modified and well-developed insects, co-existent with others of more primitive type

Further discoveries will doubtless do much to eliminate these difficulties, but present knowledge is such that recent entomology helps very little, and the classification of Palæozoic insects must be largely based on the known fossils, realising that many of the intermediate forms are yet unknown

The classifications of various authors vary widely, and even the broad general facts of relationship are still uncertain   The earliest attempt at a classification of fossil insects appears to have been that of Goldenberg (1873- 1877, ' Fauna Saræpontana Fossilis Die Fossilen Thiere aus der Steinkohlenformation von Saarbrucken) He recognised a new Order, Palæodictyoptera, for the inclusion of fossil insects differing in structure and the shape of the wings from living representatives of the Orders Neuroptera and Orthoptera, while also possessing characters which seem to link the two orders together   He did not define the order, but arranged it with the other orders as follows

Order —PALÆODICTYOPTERA Goldenb
        Genera – *Dictyoneura* Goldenb *Eugereon*, Dohrn, *Mainca* Dana, *Hemeristia* Dana, *Haplophlebium*, Scudder
Order —ORTHOPTERA
    Sub order —Orthoptera Pseudo Neuroptera
    Genera —*Termes* Goldenb *Termitidium*, Goldenb
    Sub order —Orthoptera vera
    Genus —*Blattina* German
Order —RHYNCHOTA
    Sub order —Homoptera
    Genus —*Fulgorina* Goldenb

Scudder (1887, article ' Insecta,' 'Traité de Palæontologie,' by Karl von Zittel translated by Dr Charles Barrois, vol ii Palæozoologie,' pp 716—833), in his latest classification, considerably extended that of Goldenberg, while retaining the primitive group of the Palæodictyoptera   His complete arrangement is as follows

A —PALÆODICTYOPTERA, Goldenberg
    ORTHOPTEROIDEA, Scudder
    Family —Palæoblattariæ Scudder
        Sub family —Mylacridæ Scudder
            Genera —*Mylacris* Scd , *Promylacris* Scd *Paromylacris* Scd *Lithomylacris* Scd , *Necymylacris*, Scd
        Sub family —Blattinariæ Scudder
            Genera —*Etoblattina* Scd *Spiloblattina*, Scd *Archimylacris* Scd *Anthraco-blattina*, Scd , *Gerablattina* Scd , *Hermatoblattina*, Scd , *Progonoblattina* Scd *Oryctoblattina* Scd , *Petrablattina* Scd *Poroblattina* Scd
    Family —Protophasmida Brong
        Genera —*Titanophasma*, Brong , *Lithomna* Scd *Dictyoneura*, Goldenb , *Palæophasis*, Scd , *Archæoptilus*, Scd *Protophasma*, Brong *Breyeria* de Borre, *Megarcira*, Brong *Adiophasma* Scd *Goldenbergia* Scd , *Haplophlebium*, Scd , *Paolia*, Smith , *Archæopsyllus*, Scd

Neuropteroidea Scudder
    Family – Palephemeridæ, Scd
        Genus –*Platephemera*, Scd
    Family —Homothetidæ, Scd
        Genera —*Acridites*, And *Euconus*, Scd , *Genopteryx* Scd , *Gonentomum*, Scd ,
            Didymophleps, Scd , *Homothetus* Scd , *Micoternes* Sterzel , *Omalia*,
            van Ben
    Family —Palæopterina, Scd
        Genera —*Mumia* Scd , *Diaconema* Scd *Strephoclados*, Scd
    Family —Xenoneuridæ Scd
        Genus —*Xenoneura*, Scd
    Family - Hemeristina, Scd
        Genera —*Lithomantis*, Woodw , *Lithosialis*, Scd *Pachytylopsis*, de Borre ,
            Lithentomum, Scd , *Hemeristia* Dana
    Family —Gerarina, Scd
        Genera —*Polyernus*, Scd , *Gerarus*, Scd *Adiphlebia*, Scd *Megathentomum* Scd
  Hemipteroidea, Scudder
        Genera —*Eugereon*, Dohrn , *Fulgorina*, Goldenb
  Coleopteroidea, Scudder
    Palæodictyoptera having a coleopterous aspect indicated by Geinitz and Brongniart

B —HETEROMETABOLA Packard
  Orthoptera Olivier
    Family —Forficulariæ , Latreille
    Family —Blattariæ Latreille
        Genera —*Acanthroblattina* Scd , *Sentinoblattina* Scd , *Blattidium*, Westw
            Mesoblattina, Gein
    Family —Mantidæ, Latreille
        Genus —*Mantis* Linné
    Family —Phasmidæ Leach
        Genera —*Agathemera* Scd , *Pseudoperla*, Pictet
    Families —Acridii, Latreille , Locustidæ, Latreille , Gryllidæ, Latreille
  Neuroptera, Linné
    Sub-order —Pseudoneuroptera, Erichson
    Sub order —Neuroptera vera
  Hemiptera, Linné
    Homoptera, Latreille
    Heteroptera, Latreille
  Coleoptera

C —METABOLA, Packard
  Diptera
  Lepidoptera
  Hymenoptera

A modification of this classification was introduced by Handlirsch in his
"Sub-phylum Insecta" in Eastman's translation of Zittel's 'Text-book of
Palæontology,' 1913 as follows

Class I —PTERYGOGENEA, Brauer

Order —PALÆODICTYOPTERA Goldenberg
  Families —Dictyoneuridæ, Megaptilidæ Hypermegethidæ Lithomantidæ Heolidæ
    Fouquendæ, Spilapteridæ Lamproptilidæ Polycreagridæ, Paolidæ
Order —MIXOTERMITOIDEA Handlirsch
Order —REGULOIDEA, Handlirsch
Order —PROTORTHOPTERA, Handlirsch
  Families —Spamoderidæ Ischnoneuridæ, Caloneuridæ, Sthenaropodidæ Œdischiidæ
    Geraridæ, Cacurgidæ
Order —ORTHOPTERA Olivier
  Sub-order —Locustodea Leach
  Sub-order —Acridodea Handl
Order —PHASMOIDEA, Leach
Order —DERMAPTERA De Geer
Order —DIPLOGLOSSATA De Saussure
Order —THYSANOPTERA Haliday
Order —PROTOBLATTOIDEA, Handlirsch
  Families —Stenoneuridæ, Protophasmidæ Eoblattidæ Oryctoblattinidæ Ætophlebidæ
    Cheliphlebidæ Eucaenidæ
Order —BLATTOIDEA Handlirsch
  Families —Spiloblattinidæ, Mylacridæ, Poroblattinidæ, Neorthroblattinidæ Mesoblatt-
    inidæ Pseudomylacridæ Dictyomylacridæ Neomylacridæ, Pteridomy-
    lacridæ Idiomylacridæ Dacheoblattinidæ Proteremidæ
Order —MANTOIDEA Handlirsch
      Genus —*Palæomantis* Bolton
Order —SYPHAROPTEROIDEA, Handlirsch
Order —HAPALOPTEROIDEA Handlirsch
Order —PROTOEPHEMEROIDEA, Handlirsch
Order —PROTODONATA Brongniart
Order —MEGASECOPTERA, Brongniart
Order —PROTOHEMIPTERA, Handlirsch

The publication of Handlirsch's great work, Die Fossilen Insekten, 1906—
1908, marked an important phase in the history of the study of fossil insects
Handlirsch surveyed the whole field of fossil entomology, and brought the great
bulk of the known forms under a broad classification The Order Palæodictyoptera
was much extended, defined and made to include a large series of families,
several of which, however, are clearly widely divergent in type This
was soon recognised by other workers, as doubtless by Handlirsch himself,
who may have considered it wiser to extend Goldenberg's order, even to
the inclusion of forms not definitely related, rather than to formulate a new
classification the components of which could not be rigidly defined Knowing
that the field of research was rapidly widening, Handlirsch exercised a wise
restraint in not adding a new classification, which could only be of a temporary
character Subsequent events have proved the wisdom of his action Since
1908, the study of fossil insects has attracted more students, new localities and

insect-horizons have been found, and many new types recorded Some of these linking forms already known, and others indicate relationships not fully understood The retention of the Order Palæodictyoptera has therefore resulted in the formation of a somewhat heterogeneous assemblage, all members of which have one point of agreement They are primitive co-types, more nearly related to each other in various ways than they are to recent insects, although that relationship is not always as demonstrable as one could wish

The most ambitious classification yet attempted is that of Prof Lameere (1917, 'Bull Mus Hist Nat, Paris,' no 1), who has published only a summary of his conclusions We are not able to determine how valid are his arguments, or if he is justified by evidence in setting forth his new scheme He sweeps the Order Palæodictyoptera wholly away pointing out that it consists of a heterogeneous assemblage, and substitutes a more detailed classification as follows

SUBULICORNIA

 EPHEMEROPTERA

   Family —Spilapteridæ

    Genera —*Lampropila*, Brong *Epitithe* Handl, *Becquerelia* Brong *Palæoptilus* Brong *Compsoneura*, Brong, *Spilaptilus*, Handl *Homaloneura* Brong, *Graphoptilus*, Brong *Spilaptera* Brong

   Family —Megasecopteridæ

    Genera —*Aspidothorax*, Brong, *Corydaloides*, Brong, *Diaphanoptera*, Brong, *Cyclosialis*, Brong *Sphecoptera*, Brong, *Psilothorax*, Brong, *Mischoptera*, Brong, *Ischnoptilus*, Brong

   Family —Protephemeridæ

    Genera —*Appappus*, Handl, *Triplosoba* Handl

 ODONATOPTERA

   Family —Fouqueidæ

    Genera —*Fouquea*, Brong, *Rhabdoptilus* Brong

   Family —Dictyoneuridæ

    Genera —*Macrodictya*, Brong, *Stenodictya*, Brong

   Family —Dictyoptilidæ (Protodonata)

    Genera —*Archæmegaptilus*, Brong *Dictyoptilus*, Brong, *Poromaptera*, Brong *Protagrion*, Brong *Gilsonia* Brong, *Meganeura*, Brong

RHYNCHOTA

PROTOHEMIPTERA

   Family —Homoropteridæ

    Genera —*Lycocercus*, Handl, *Homoroptera*, Brong *Lithoptilus* Brong

   Family —Megaptilidæ

    Genus —*Megaptilus*, Brong

   Family —Breyeridæ

    Genus —*Megaptiloides*, Handl, *Breyeria* Brong

   Family —Mecynostomidæ

    Genus —*Mecynostoma*, Brong

HEMIPTERA

   Family —Dictyocicadidæ

    Genus —*Dictyocicada*, Brong

Class I.—PTERYGOGENEA, Brauer

   Order.—PALÆODICTYOPTERA, Goldenberg
          Families.—Dictyoneuridæ, Megaptilidæ, Hypermegethidæ, Lithomantidæ, Heolidæ
              Fouqueidæ, Spilapteridæ, Lamproptilidæ, Polycreagridæ, Paolidæ
   Order.—MIXOTERMITOIDEA, Handlirsch
   Order.—RECULOIDEA, Handlirsch
   Order.—PROTORTHOPTERA, Handlirsch
          Families.—Spanioderidæ, Ischnoneuridæ, Caloneuridæ, Sthenaropodidæ, Œdischiidæ
              Geraridæ, Cacurgidæ
   Order.—ORTHOPTERA, Olivier
      Sub-order.—Locustodea, Leach
      Sub-order.—Acrodiodea, Handl
   Order.—PHASMOIDEA, Leach
   Order.—DERMAPTERA, De Geer
   Order.—DIPLOGLOSSATA, De Saussure
   Order.—THYSANOPTERA, Haliday
   Order.—PROTOBLATTOIDEA, Handlirsch
          Families.—Stenoneuridæ, Protophasmidæ, Eoblattidæ, Oryctoblattinidæ, Ætophlebidæ
              Cheliphlebidæ, Eucænidæ
   Order.—BLATTOIDEA, Handlirsch
          Families.—Spiloblattinidæ, Mylacridæ, Poroblattinidæ, Neorthroblattinidæ, Mesoblat-
              tinidæ, Pseudomylacridæ, Dictyomylacridæ, Neomylacridæ, Pteridomy-
              lacridæ, Idiomylacridæ, Diechoblattinidæ, Proteremidæ
   Order.—MANTOIDEA, Handlirsch
             Genus.—*Palæomantis* Bolton
   Order.—SYPHAROPTEROIDEA, Handlirsch
   Order.—HAPALOPTEROIDEA, Handlirsch
   Order.—PROTEPHEMEROIDEA, Handlirsch
   Order.—PROTODONATA, Brongniart
   Order.—MEGASECOPTERA, Brongniart
   Order.—PROTOHEMIPTERA, Handlirsch

The publication of Handlirsch's great work, Die Fossilen Insekten, 1906—
1908, marked an important phase in the history of the study of fossil insects.
Handlirsch surveyed the whole field of fossil entomology, and brought the great
bulk of the known forms under a broad classification. The Order Palæodictyoptera
was much extended, defined and made to include a large series of families,
several of which, however, are clearly widely divergent in type. This
was soon recognised by other workers, as doubtless by Handlirsch himself,
who may have considered it wiser to extend Goldenberg's order, even to
the inclusion of forms not definitely related, rather than to formulate a new
classification the components of which could not be rigidly defined. Knowing
that the field of research was rapidly widening, Handlirsch exercised a wise
restraint in not adding a new classification, which could only be of a temporary
character. Subsequent events have proved the wisdom of his action. Since
1908, the study of fossil insects has attracted more students, new localities and

insect-horizons have been found, and many new types recorded Some of these linking forms already known, and others indicate relationships not fully understood The retention of the Order Palaeodictyoptera has therefore resulted in the formation of a somewhat heterogeneous assemblage, all members of which have one point of agreement They are primitive co-types, more nearly related to each other in various ways than they are to recent insects, although that relationship is not always as demonstrable as one could wish

The most ambitious classification yet attempted is that of Prof Lameere (1917, 'Bull Mus Hist Nat , Paris,' no 1), who has published only a summary of his conclusions We are not able to determine how valid are his arguments, or if he is justified by evidence in setting forth his new scheme He sweeps the Order Palaeodictyoptera wholly away pointing out that it consists of a heterogeneous assemblage, and substitutes a more detailed classification as follows

SUBULICORNIA
    EPHEMEROPTERA
        Family —Spilapteridæ
           Genera — *Lamproptila*, Brong *Eplithe* Handl , *Becquerelia* Brong *Palœoptilus* Brong *Compsoneura*, Brong , *Spiloptilus*, Handl *Homaloneura* Brong , *Graphoptilus*, Brong *Spilaptera* Brong
        Family —Megasecopteridæ
           Genera —*Aspidothorax*, Brong , *Corydaloides*, Brong , *Diaphanoptera*, Brong , *Cycloscelis*, Brong *Sphecoptera*, Brong , *Psilothorax*, Brong , *Mischoptera*, Brong , *Ischnoptilus*, Brong
        Family —Protephemeridæ
           Genera —*Appapapus*, Handl , *Triplosoba*, Handl
    ODONATOPTERA
        Family —Fouqueidæ
           Genera —*Fouquea*, Brong , *Rhabdoptilus* Brong
        Family —Dictyoneuridæ
           Genera —*Microdictya*, Brong , *Stenodictya*, Brong
        Family —Dictyoptilidæ (Protodonata)
           Genera —*Archœmegaptilus*, Brong *Dictyoptilus*, Brong , *Paronaptera*, Brong *Protagrion*, Brong *Gilsonia* Brong , *Meganeura*, Brong

RHYNCHOTA
    PROTOHEMIPTERA
        Family — Homopteridæ
           Genera —*Lycocercus*, Handl , *Homoioptera*, Brong *Lithoptilus* Brong
        Family —Megaptilidæ
           Genus —*Megaptilus*, Brong
        Family — Breyeridæ
           Genus —*Megaptiloides*, Handl , *Breyeria* Brong
        Family —Mecynostomidæ
           Genus —*Mecynostoma*, Brong
    HEMIPTERA
        Family —Dictyocicadidæ
           Genus —*Dictyocicada*, Brong

ORTHOPTERA
   Nomoneura (Blattaeformia Handl.)
     (a) Blattoidea
        Families —Hyaloptilidæ, Protoperlidæ, Fayolidæ, Oryctoblattinida
     (b) Mantoidea
        Families —Stenoneuritidæ, Stenoneuridæ, Ischnoneuridæ
   Heteroneura (Equivalent in part to Orthoptera Cursoria and Orthopteroidea Handl.)
     (a) Plasmodea
        Family —Sthenaropodidæ
     (b) Locustidæ (Equivalent in part to Orthoptera Saltatoria.)
        Families —Œdischiidæ, Caloneurida

Lameere restricted his research to the French fossil insects, his studies being based on the types described and figured by Brongniart (1893, 'Bull. Soc. Industrie, Saint Étienne, 3 sér., VII), the collections made by Fayol, and the large series of fossil insects from the Upper Coal Measures (Stephanian) of Commentry now preserved in the National Museum of Natural History, Paris. The following remarks may be made on his classification.

EPHEMEROPTERA.—The three families forming this division are regarded as closely related, the genus *Becquerelia* of the Spilapteridæ bearing certain characters of the Megasecopteridæ, while the family is also linked through the genus *Apopappus* (which is taken to supply a natural transition between the Spilapteridæ) to the genus *Triplosoba*.

ODONATOPTERA.—The family Fouqueidæ is held to differ from the Spilapteridæ in that transverse veins are numerous, close together, and form a network over the inner margin, and in the anal area—a feature which brings it nearer to the Protodonata. The family Dictyoneuridæ possesses a network of veins extending over the whole wing as in *Microdictya*. The remaining family, Dictyoptilidæ, contains *Archæmegaptilus* which Lameere considers differs only from the Dictyoneuridæ in the fusion of the median and radial veins at the base of the wing. *Protagrion* is considered nearest to the true Odonata, while *Meganeura* and *Gilsonia* are specialised types.

RHYNCHOTA.—The presence of a rostrum in *Lycocercus goldenbergi*, and the resemblance of the head and leg of *Homoioptera gigantea* to the same structures in *Eugereon*, are considered sufficient proof of the Protohemipteroid characters of *Lycocercus Homoioptera* and the allied genera *Lithoptilus*, *Megaptilus*, *Mecynostoma Archæoptilus* and *Paramegaptilus*.

ORTHOPTERA.—Lameere regards Handlirsch's group of Protorthoptera as an assemblage of two related but distinct types which he classifies under Nomoneura and Heteroneura. The genus *Stenoneurites* is regarded as the connecting link between the Mantoidea and the ancestors of the Blattoidea, the genus *Stenoneura* being also in some measure transitional between *Stenoneurites* and the Ischnoneuridæ.

Nomoneura.—This sub-division includes the Blattæformia of Handlirsch and is distinguished by the wings having no precostal area, as contrasted with a second division, Heteroneura, in which a precostal area is present, and in which the legs are adapted for running and jumping. The Nomoneura include forms classified by Handlirsch under the Protorthoptera and Protoblattoidea. Lameere separates his Nomoneura into (*a*) Blattoidea, and (*b*) Mantoidea, the former characterised by a sub-costal which joins the outer or costal margin, a more or less lengthened radius, and a small cubitus. The Mantoidea have the sub-costa joining the radius, and a large preponderating cubitus.

Heteroneura.—This sub-division includes the Phasmoidea and the Locustoidea, the former containing the family (*a*) Sthenaropodidæ in which the legs are long and stout, the head prognathous, the prothorax long and narrow in front and very wide behind, and presenting two dorsal expansions.

The wing-venation of the Sthenaropodidæ is such that they may have been the ancestors of the Phasmidæ. Lameere, however, does not regard the Sthenaropodidæ as ancestors of the Phasmidæ, but as arising with them from a common ancestor.

The members of the families Œdischiidæ and Caloneuridæ possess legs fitted for jumping, but differ considerably in their wing-venation. The Œdischiidæ are possibly linked with the Locustidæ, and the Caloneuridæ with the Acrididæ.

Circumstantial and detailed as Prof. Lameere's classification is, the arguments and deductions are not easily followed, the paper being only in abstract. A study of the fossil insect-material from Commentry alone is not in itself likely to yield all the facts and premisses upon which a classification can be built applicable to the Palæozoic insect-fauna of all coalfields and countries. Much more evidence is wanted, and until the full paper is published, it is necessary to hold the classification in suspense.

The most recent publication on the Palæozoic insects is an extensive and valuable memoir by Dr. P. Pruvost (1920, "La Faune Continentale du Terrain Houiller du Nord de la France," 'Mémoires pour servir à l'Explication de la Carte Géologique détaillée de la France,' Paris, 1920) on the fossil insects recently found by him, and others, in the neighbourhood of Lens and Liévin in the north of France. Dr. Pruvost adopts the classification of Handlirsch with few emendations as follows, and by his new material he has added considerably to our knowledge of the Protoblattoidea and Blattoidea.

Order —PALÆODICTYOPTERA
　　　　Family —Stenodictyopteridæ, Brong. (Dictyoneuridæ, Handlirsch)
　　　　Family —Spilapteridæ, Brong. emend. Handlirsch
Order —PROTORTHOPTERA, Handlirsch
　　　　Families —Œdischiidæ, Caloneuridæ
Order —HAPALOPTEROIDEA, Handlirsch
　　　　Family —Hapalopteridæ

3

Order—PROTOBLATTOIDEA Handlirsch emend Pruvost.
    Suborder—Archoblattodea, Pruvost.
        Family—Chonodea Handl
    Suborder—Archmantodea Pruvost
        Family—Cymenophlebidea, Pruvost
Order—BLATTOIDEA Handlirsch
    Family—Archoblattidae, Handl
        Genera—*Archoblattina* *Anomoblatta* *Manoblatta, Actinoblatta* *Phyloblatta* *Archeo
            tipia* *Petrosoblatta* *Gerypoblatta* *Mesoblattia*
    Family—Mylacridae Scd
        Genera—*Hermatylacris* *Phylomylacris, Tritophomylacris Soomylacris Orthomylacris
            Stenomylacris and Lithomylacris*
    Family—Poroblattinidae Handl
        Genus—*Premnoblatta* Prvs

I have compared notes with M. Pruvost and we have arrived independently
at the conclusion that for the present the classification of Handlirsch is, with few
emendations, the best to adopt, and most in keeping with the known facts

FAUNAL ASSOCIATION

Various collectors in British Coalfields have discovered not only insect-remains,
but a definite faunal association, of which the significance seems to have been
overlooked, and it has therefore not received the attention it deserves

Most of the insect-remains are found in ironstone nodules, and the beds in
which these nodules occur are usually light-coloured rocks more similar to hardened
clay than to normal shales. The nodules are in vast numbers, ranging in size from
half-an-inch to ten and twelve inches in diameter. The beds seem more comparable to the fireclays or seat-earths than to the ordinary fissile shales, and both
in lithological character and fossil contents stand in some measure apart from the
ordinary Coal Measure rocks. They are not restricted to one coalfield, but have a
wide distribution. Where a systematic search of beds of this character has been
made, the insect-remains have been found accompanied by a fauna in which
arthropods of a more primitive type than insects are conspicuous

The character of this fauna will be best understood by reference to the
following lists of fossils which have been recorded from certain localities

DURHAM COALFIELD—' Zone of *Anthraconia phillipsii* (Will.) ' in upper part
of the Middle Coal Measures. Claxheugh escarpment two miles west of Sunderland, Durham

PELECYPODA	MEROSTOMATA
' *Anodus indi,* Kirkby " (cf. ' Spit' of	*Belinurus trochmanni* (Woodw.)
Anthraconia phillipsii, Will.)	DIPLOPODA
Anthraconia minima (Ludvig)	*Euphoberia* sp
var. *scotica*	

OSTRACODA
  Beyrichia McCoy
  Cythere or *Cypris*

INSECTA
  Lithomylacris kirkbyi, Woodward
  Phyloblattia mantidioides (Goldenberg)
PISCES
  Rhizodopsis sauroides (Will.)

NOTTINGHAM AND DERBYSHIRE COALFIELD.—Below the Top Hard Coal, Middle Coal Measures (1911, Moysey 'Geol Mag' [5] vol viii p 506), Shipley Manor clay-pit, one and a half miles north of Ilkeston, Derbyshire

ANNELIDA
  Spirorbis, sp
PELECYPODA
  Anthracomya modiolaris (Sow)
  Carbonicola aquilina (Sow)
  Naiadites modiolaris (Sow)
OSTRACODA
  Beyrichia arcuata ? (Bean)
  Estheria sp
  Leaia tricornoides, Moysey
CRUSTACEA
  Pranaspides precursor, Woodw
  Arthropleura armata Jordan
  „   sp nov , Moysey
ARACHNIDA
  Cyclus sp
      johnsoni, Woodw
  Belinurus bellulus Konig
  „   *koenigianus* Woodw
      longicaudatus, Woodw
      sp
  Prestwichia anthrax (Prestw)
  „   *butterlli*, Woodw
      rotundata (Prestw)
      sp
  Eurypterus derbiensis Woodw

ARACHNIDA—(continued)
  Eurypterus moyseyi Woodw
  Scorpion, post-abdominal segment, Moysey
  Geralinura britannica, Pocock
  Eophrynus hulti, Pocock
  Anthracosiro fritschi Pocock
  „   *woodwardi*, Pocock
  „   sp
  Protolycosa sp
INSECTA
  Cryptovenia moyseyi Bolton
  Orthocosta splendens, Bolton
  Pterorachis plicatula Bolton
PISCES
  Elonichthys, sp
EGG CAPSULES OF FISHES (?)
  Fayolia crenulata Moysey
      cf *dentata* Zeiller
  Palaeoxyris carbonarius (Schimper)
      helictoroides (Morris)
      prendeli, Lesq
  Vetacapsula johnsoni (Kidston)
  „   *cooperi*, Mackie and Crocker
PLANT REMAINS
  Annularia , Calamocladus Sphenophyllum ,
      Lepidophyllum , Calamites ferns

LANCASHIRE COALFIELD.—Greyish-blue shales, 135—180 feet above the Royley or Arley mine, the latter at the base of the Middle Coal Measures, Spath Bottoms, Rochdale Lancashire   I am indebted for the following list to Mr Walter Baldwin, F G S , who, with Messrs Sutcliffe, Parker, Platt and others, has devoted years to the examination of these beds

VERMES
  Spirorbis (Spirorbulus)
BRACHIOPODA
  Lingula, sp

PELECYPODA
  Carbonicola acuta (Sow)
      robusta (Sow)
  „   *turgida* (Brown)

PELECYPODA—(continued)
  Naiadites modiolaris (Sow.)
             triangularis (Sow.)
             carinata (Sow.)
             elongata (Hind)
  Anthracomya williamsoni (Brown)

EUCRUSTACEA
  Dithyrocaris sp.
  Prjapephalus cooperi Huxley
            (? Anthrapalaemon) parkeri
                                (Woodw.)

  Anthrapalaemon grossarti Salter
  Eurypterus, sp.
  Cyclus johnsoni, Woodw.
  Rochdalia parkeri Woodw.
  Belinurus lunatus (Martin)
      „   lusitanicus, Woodw.
      „   baldwini Woodw.
      „   longicaudatus Woodw.
      „   testudineus (Woodw.)

ARACHNIDA
  Prestwichia birtwelli Woodw.
      „   rotundata (Prestwich)
      „   var. major, Woodw.
                    minor Woodw.
          (Euproöps) dawsoni Meek & Worthen
          anthrax (Prestwich)
  Eoscorpius (Mazonia) woodwardi Woodw.
  Lobolthus holti Pocock

ARACHNIDA—(continued)
  Anthracoscorpio bathytormis Pocock
            „   spuethensis Pocock
  Geralinura scutelifer Woodw.
  Anthracomartus triloblatus Scudder
      „   sp. 1 in Platt Coll.
      „   sp. 2 in Platt Coll.
  Anthracosro woodwardi, Pocock
  Phalangiotarbus subovalis (Woodw.)
          (Architarbus) rotundata
                            Woodw.

DIPLOPODA
  Xylobius platti Woodw.
  Euphoberia ferox Salter
      „   armigera (Baldwin)
      „   robusta (Baldwin)
      „   woodwardi (Baldwin)
  Archiulus sp.
  Acantherpestes major, Meek & Worthen
            gigateus, Baldwin

INSECTA
  Spilaptera sutcliffe Bolton
  Megaptilus tuberculata, Bolton

PISCES
  Platysomus tenuistriatus, Traquair
  Rhizodopsis sauroides (Will.)

INCERTE SEDIS
  Palaeoxyris prendeli Lesq.

SOUTH STAFFORDSHIRE COALFIELD.—Binds between the Brooch and Thick coals, Tipton, Dudley and Coseley

EUCRUSTACEA
  Euphoberia ferox Salter
ARACHNIDA
  Anthracoscorpio bathytormis Pocock
  Geralinura britannica Pocock
  Geraphrynus anglicus Pocock
  Eocteniza silvicola Pocock
  Archaeometa nephalina, Pocock
  Curculioides austeni Buckland
  Polyochera alticeps Pocock
  Plesiosiro madeleyi Pocock
  Geraphrynus anglicus Pocock
      „   hindi Pocock
      „   tuberculatus, Pocock
      „   oppatoni Pocock
      „   torpedo, Pocock

ARACHNIDA—(continued)
  Geraphrynus angustus, Pocock
  Anthracomartus hindi Pocock
            priesti Pocock
  Anthracosro woodwardi Pocock
            tutschi Pocock
  Eoguodartbus johnsoni Pocock

INSECTA
  Palaeodictyoptera
    Pearstia spiculatus Bolton
    Brodia priseutincta Scudder
      „   furcata (Hand.)
    Pleronequonides hrpus, Bolton
            ambiguus Bolton
    Germaria (?) ovata Bolton

INSECTA—(continued)
  Protorthoptera
    Aëropteria obtusata, Bolton
    Sialeoptera recta, Bolton
    Coselia palmiformis, Bolton
  Blattoidea
    Aphthoroblattina johnsoni (Woodw.)

INSECTA—(continued)
  Blattoidea—(continued)
    Aphthoroblattina eggintoni (Bolton)
    Archimylacris incisa (Bolton)
    Phylloblatta transversalis (Bolton)
  LARVAL BLATTOID
    Leptoblattina exilis, Woodw.

COALBROOKDALE COALFIELD, SHROPSHIRE.—Ironstone Nodules of the Pennystone Series

EUCRUSTACEA
  Euphoberia ferox, Salter
  Anthrapalæmon (Apus) dubius (Prestw.)
ARACHNIDA
  Prestwichia anthrax (Prestw.)
    „   rotundata (Prestw.)
    „   trilobitoides, Woodw.

ARACHNIDA—(continued)
  Curculioides ansticii, Buckland
  Eophrynus prestvici (Buckland)
DIPTEROIDA
  Acantherpestes brodiei, Scl.
INSECTA
  Lithosialis brongniarti (Mantell)

SOUTH WALES COALFIELD.—Shales in the neighbourhood of the Mynddslwyn Vein, base of the Upper Coal Measures

ARACHNIDA
  Aphantomartus areolatus Pocock
  Graeophonus anglicus Pocock
  Mancerus cellicus Pocock
  Kreischeria verrucosa, Pocock

INSECTA
  Aphthoroblattina sulcata (Bolton)
  Orthomylacris lanceolata (Bolton)
  Archimylacris hastata (Bolton)
    „   obovata (Bolton)

The arthropod association in the lists given is significant, for no other animals enumerated are so readily water-borne as are insects. It may be assumed that neither the more primitive arthropods, nor the insects, have been transported to any great distance from their former habitat. Their preservation under similar conditions supports the belief that their habits and habitats were the same, or closely approximated to the same conditions of entombment.

The freedom of the deposits from comminuted carbonaceous matter such as is usually a chief constituent of the Coal Shales may be accounted for by the beds having been laid down in quiet lagoons or swamp lakes, into which only the finer mud particles and floating pinnules and debris of coal-plants could pass and accumulate. Such waters were probably fresh or brackish, shallow, and limited in area. As we have indicated elsewhere (p. 10), it has been considered that the larvæ of many of the Coal Measure insects were semi- or wholly aquatic, and if such was the case, they would be more likely to be found in the deposits accumulating in quiet waters than in others exposed to movement. The presence of Mollusca and the lower orders of Arthropoda, with such forms as *Palæocaris*, and even fish-remains, can be accounted for by the existence of occasional or permanent passages leading to open waters, such as river channels and the open sea.

A succession of such lagunal lakes or swamp pools might be seasonal features along a depression which, in a wet season, formed a water-course

The paucity of Mollusca is noteworthy, only one species ('*Unio*') having been recorded by Moysey, notwithstanding his careful search of material from the Shipley clay-pit at Ilkeston, Derbyshire while Kirkby, and more recently Trechmann and Woolacott have recorded *Anthracomya phillipsii*, in addition to '*Anodus rutti*' (now known to be the larval form of *A. phillipsii*), from the insect-bearing beds at Claxheugh, near Sunderland

The deposits at Sparth Bottoms Rochdale, Lancashire are remarkable in that they have yielded three species of *Carbonicola* four species of *Naiadites*, and one species of *Anthracomya*

In regard to the presence of Crustacea, Thomson (1894, 'Trans Linn Soc London, Zoology [2], vol vi, p 3), has shown that the recent *Anaspides tasmaniae* lives in freshwater pools and lakes which are wholly cut off from the sea, and Dr H. Woodward (1908, 'Geol Mag' [5] vol v p 385) has described an allied form, *Pranaspides precursor*, from the Coal Measures of Shipley Clay-pit Further research may prove that not only *Pranaspides*, but such forms as *Belinurus*, *Prestwichia Eurypterus* and *Anthrapalamon* were also fresh-water in habit, co-existing with insect larvæ in the lagoons and swamp lakes

Observations by the writer during the visit of the British Association to Australia in 1914 bear on this point While collecting in the "Bush country" at St Margaret's Bay, Western Australia, at Warburton, S Australia, and elsewhere, examination of almost every loose sheet of bark hanging to the gum trees revealed a colony of scorpions, millipedes, spiders and cockroaches In the Australian localities mentioned true "bush" conditions prevailed, and seemed much unlike those of low-lying swamps, such as are predicated for the Coal Measure period Subsequent experiences along the coast of North Queensland modified these first impressions It was found that notwithstanding the hot tropical day, or perhaps by reason of it, the nights brought an extremely heavy dew, so that it was impossible to move ten yards in the jungle before the clothing was running with water discharged from the leaves of the jungle plants In a short time after sunrise the jungle was dry again but much of the moisture must have been caught up under projections capable of resisting the penetration of the sun's heat In the "bush" country, slabs of bark may give conditions which the arthropod fauna find the most tolerable during the hot season The Australian winter with its heavy rains would more nearly accord with swamp conditions Long stretches of the coastline of Queensland, north of Brisbane, the country along the Hinchinbrooke Channel, and in the neighbourhood of Townsville, are low-lying, and covered by dense mangrove swamps growing out into the sea and completely hiding the outlets of rivers

The conditions seem similar to those of the low-level country, swamp and mud flat upon which the coal forests are believed to have flourished. During the voyage along the coast a landing was made at Lucinda Point, and an opportunity was afforded of entering the swamp. Part of the swamp was awash on landing, but as the tide receded, it was possible to reach the shore. The latter was found to be a flat shelving beach consisting of fine marine sand on the seaward side, passing shorewards into a fine tenacious clayey mud in which the mangroves grew. The sand bore numerous remains of echini, mollusca and marine *debris*, while the mud below high-water mark and above it was penetrated in all directions by the roots of mangroves, by the burrows of crabs, and by a burrowing gasteropod, *Telescopium fuscum*.

Masses of leaves, branches and other vegetable material were mixed with the mud, the latter in some places forming irregular lumps and boulder-like masses around the vegetable material. A broad shallow depression, with a few inches of water over a thick bed of mud littered with leaves, led into the swamp. It was evidently the bed of a stream during the wet season, remaining as a series of disconnected pools at other times, or drying up.

The resemblance of the physical conditions to those accepted as dominant in the Coal Measure period was so evident that special attention was given to it with a view to discovering discrepancies; yet had the mangroves been replaced by dense groves of gum trees with their arthropod fauna, the circumstances would have been well-nigh identical. Had the faunal association of scorpions millipedes spiders and cockroaches of West Australia been transported to the swamp, their remains would have become entombed in material not unlike that at Sparth Bottoms or Coseley.

There is ample evidence to show that the coal plants grew on ground of this character, and we know that the remains of fossil insects are associated with the plant-remains.

The swamp conditions of the North Queensland coast reproduce with great exactitude the presumed Coal Measure conditions, but lack the arthropod fauna owing to the unsuitability of the vegetation.

Scudder had some such habitat and faunal association in mind when writing his work "Archipolypoda a Sub-ordinal Type of Spined Myriapods from the Carboniferous Formation" (Mem. Boston Soc. Nat. Hist.' vol. iii, no. 5, pp. 143—182). Plate X of that work is described as "an attempted restoration of a specimen of *Acantherpestes major*, Scd.' The specimen is represented as leaving the water, in which the hinder part of the body is still swimming by means of its legs, while the fore part of the body is creeping up the trunk of a Lepidodendron (*L. vestitum*). On the trunk crawls a cockroach (*Etoblattina mazona* Scd.), and a broken stem of *Calamites cistii*, Brong, lies partly fallen in a clump of *Neuropteris loschii*, Lesq.

We are, therefore, not without some justification in assuming that the faunal association which exists at Coseley, Sparth Bottoms and elsewhere is not due to the accident of deposition or transportation, but the natural result of the conditions under which insect life was passed

SYSTEMATIC DESCRIPTIONS

Order *PALEODICTYOPTERA* (Goldenberg), Handlirsch

1877 Goldenberg, Die Fossilen Thiere aus der Steinkohlenformation von Saarbrucken   Fauna Saraepontana Fossilis pt ii p 8
1906 Handlirsch Die Fossilen Insekten p 61

Slender insects with moderate-sized head and biting jaws simple antennæ, and two pairs of equal and similarly shaped wings   The wing-venation is not unlike the hypothetical tracheation of the primitive nymph worked out by Comstock and Needham (American Naturalist,' vol xxxii, no 374, p 85, fig 4. 1898—99)

The wings could not be folded, being outstretched laterally in the position of rest, and only moving in a vertical plane at right-angles to the body   The thoracic segments are three in number, with wing-like pleurites in some cases on the first segment   The abdomen has eleven segments the eleventh segment bearing cerci   Legs all similar and fitted for running

Goldenberg did not define the characters of the Order, but included in it a group of Palæozoic insects which, while somewhat related to the existing forms of Nemoptera, are sufficiently unlike to prevent their inclusion in the latter Order

Handlirsch defined the Order as a primitive generalised group, probably the ancestors of all later insects, and wholly confined to the Palæozoic   He considered that the larvæ were predatory and aquatic developing their wings gradually without resting stages, and being in other respects similar to the imago   His diagram of the primitive Palæodictyopterous wing shows a slight advance upon the primitive nymph of Comstock and Needham, the primary tracheation being increased by the development of cross-nervures, uniting to form a meshwork

Family DICTYONEURIDÆ, Handlirsch

1906 Handlirsch Die Fossilen Insekten, p 63
1919 Handlirsch Revision der Paläozoischen Insekten p 3

Palæozoic insects in which the wings possess a close reticulated neuration between the principal veins, the latter strong and parallel over the first third of the wing   Branches of the radial sector, median and cubitus few in number, and strongly curved back to the inner margin

Handlirsch regards this family as closely related to *Mecodactya*

Genus **DICTYONEURA**, Goldenberg

1854   *Dictyoneura*, Goldenberg Palaeontographica vol iv p 33

Wing three times as long as wide   Base narrow, outer margin curving regularly backwards beyond the middle of the wing into the wing-apex   Inner margin rounded, and joining the apex in a blunt curve   Subcosta straight, reaching the costal border in the outer third of the wing   Radius simple, parallel to subcosta, and giving off a radial sector with few branches   Hinder branch of median and first branch of cubitus uniting to form a "cubito-median Y-vein"   Anal veins with few branches   Interstitial neuration irregularly reticulate

Dictyoneura higginsii (Handlirsch)   Plate I fig 1

1871   Neuropterous Insect-Wing, Higgins Proc Liverpool Nat Field Club vol ii, p 18 fig 15
1906   (*Palaeodictyopteron*) *higginsii* Handlirsch Foss Insekt , p 125, pl xiii fig 6
1917   (*Dictyoneuron*) *higginsii*, Bolton, Quart Journ Geol Soc , vol lxxiii p 46, pl iii, fig 2
1919   *Sherbornrella higginsii*, Handlirsch Revis Palaozoischen Insekten p 25

Type —Basal portion of a left wing in counterpart , Liverpool Museum

Horizon and Locality —Middle Coal Measures , Ravenhead railway cutting near St Helens, Lancashire

Description —The wing-fragment lies on the surface of a greyish-purple ironstone nodule, and has a length of 32 mm with a greatest width of 22 mm The total length and width cannot be exactly estimated

Handlirsch, who named the species, did not describe it and it is doubtful if he ever saw it   It was first described by the late Rev H H Higgins thus "A second and smaller specimen of insect wing was obtained by myself, and referred to the genus *Corydalis*   Mr J P G Smith compared it with *Fulgora* A slight sketch of it was seen by Mr F Smith, of the British Museum, whom it reminded of *Gryllotalpa*   Mr Benj Cooke, of Manchester, after a careful examination says 'I believe the fossil represents the basal portion about one-third only, of the forewing of a *Chrysopa*, Goldeneye, or Lace-wing fly or rather *Nothochrysa*, separated from *Chrysopa* by Mr McLachlan on account of the manner in which the third cubital cell is divided'   This cell is remarkably well shown in the fossil and though I could only judge from memory, I believe it is sufficient to settle its relationship"

The 'cubital cell   mentioned by Mr. Benj Cooke is an elongated area enclosed by a union of two of the main veins in the base of the wing   The identity of the veins enclosing the "cell' will be considered later

The costal border is preserved for a length of 29 mm , is strongly curved proximally and less so distally   It would seem to have been directed somewhat

4

backwards in its course to the wing-apex. The intercostal area is very wide at the base, and diminishes in diameter towards the wing-apex, but it is doubtful if it reached the latter.

The subcosta is straight, but not well defined. The radius is fairly parallel to the subcosta and gives off the radial sector low down in the base of the wing; the radial sector comes off from the radius at an acute angle, and keeps parallel with it as far as preserved. Its direction is such that it must have reached the wing-apex. The median and cubital veins merit special consideration. In the specimen three veins occupy the position of the normal median and cubitus. Of these, two either arise from a common root, or so close together as to be indistinguishable; the third having a separate origin. The innermost of the three sends off from near its base a forwardly directed twig which joins the second of the two outer veins.

It is necessary to resolve these veins into median and cubitus. To do this we must consider recent research on living insects. Comstock and Needham concluded that the primitive median vein was a four-branched structure; more recently Tillyard, from further studies of the wing-venation of recent nymphs, concludes that the primitive median had an initial dichotomy, of which the outer branch divided up into four ('M' of Tillyard 'Proc. Linn. Soc. N.S. Wales' vol. xliv. 1919, p. 552), and the second branch remained simple. Comstock and Needham concluded that the primitive cubitus was two-branched. Tillyard considers it (loc. cit., p. 553) as three-branched, the outer or first branch having a distal forking into two feeble twigs, and an inner feeble branch which remains simple. The primary cubital fork is situated near the base of the wing.

Tillyard's views are not very different from those of Comstock and Needham, as he admits a basal fork of the cubitus but adds a secondary forking of the end of the outer branch.

Tillyard has studied the relation of the two veins in the various Orders, and finds that the "posterior arculus" of Comstock, which is supposed to be a cross-vein from the median to the cubitus, is, as shown in the fossil Order Paramecoptera, not a cross-vein, but a true branch of the median. The various stages by which the connection becomes established and afterwards developed into a combined vein from the point of union are fully stated (loc. cit., p. 557), and the compound vein is named the "cubito-median Y-vein." An examination of the conditions observable in *Dictyoneura higginsi* shows that the united veins and their single-stemmed prolongation are in position and character identical with Tillyard's "cubito-median Y-vein." The three veins of the specimen, therefore, are the outer median vein, the inner median uniting with the first cubital vein to form a "cubito-median Y-vein," and lastly, the second (inner) branch of the cubitus.

Tillyard recognises the presence of the "cubito-median Y-vein" in the Permian Order Paramecoptera, but its discovery in the Coal Measure genus *Dictyoneura* shows that its origin is of much older date. The cubito-median Y-vein lies close to the outer median for a short distance, and then bends obliquely inwards to the margin of the wing. The inner cubital vein is fairly straight, and widely spaced along almost its whole length from the cubito-median Y-vein.

Six anal veins are present, all going obliquely to the inner margin, the second and third alone forking. The interstitial neuration, which I have been able to study better by immersing the specimen in water, consists of numerous irregular cross-nervures, occasionally uniting in crossing, or forming a loose meshwork, especially between the median and cubital veins.

Affinities.—With so small a wing fragment, and with few wings of similar character as guides, it is not possible to form an accurate idea of the whole wing. The radius and radial sector entered the wing-apex, and the median-cubital elements occupied the greater part of the inner margin of the wing.

The broad intercostal space and the course of the subcosta are very similar to those in *Palaeoagla elegans*, Handlirsch, but the general trend of the main veins, and most of all the character of the interstitial neuration, are so suggestive of *Dictyoneura libelluloides*, Goldenberg, that I had referred the specimen to that genus before Handlirsch's paper was published.

Handlirsch has now (1919) referred this species to a new genus, *Sherlockiella*, giving the new name 'for the sake of uniformity,' but without diagnosis.

Family ORTHOCOSTIDÆ, Bolton

1912 Bolton Quart Journ Geol Soc vol lxviii p 313

Wings with almost straight outer margin, inner margin well rounded; costa, subcosta and radius closely approximated; median, cubitus and anal veins occupying nearly the whole of the inner half of the wing.

Genus **ORTHOCOSTA**, Bolton

1912 *Orthocosta* Bolton *loc cit*, p 313

Generic Characters.—Radial sector diverging from radius, and with few divisions. Median with two branches united by a commissure. Cubitus stout, forked near base, the outer branch forking twice and the inner once. Anal area elliptical. Interstitial neuration forming an open polygonal meshwork.

Orthocosta splendens, Bolton   Plate 1 fig 2   Text-figure 1

1912 *Orthocosta splendens*, Bolton, Quart Journ Geol Soc, vol lxviii p 310, pl xxxi figs 1—3

Type.—Incomplete wing, and impression. Mowsey Collection, Museum of Practical Geology, Jermyn Street (nos 30,222 and 30 223)

Horizon and Locality —Middle Coal Measures (below the Top Hard Coal), Shipley Manor Claypit, Ilkeston, Derbyshire

Specific Characters —Radial sector reaching the inner half of the wing-tip Outer branch of median four or five times divided inner branch simpler and forked   Cubitus dividing low down the outer branch the stronger, and each doubly forked   Anal veins one or two in number, alternately branched

Description —The species is founded on a wing-fragment the apex, a portion of the inner margin, and the base being incomplete   The total length is 84 mm the width 33 mm   The complete wing must have had a length of at least 100 mm, and a width of 35—40 mm   The whole insect had in all probability a span of wing measuring 25 5 cm (10 in or more)

The outer third of the wing is differentiated from the rest by its uniform and gentle convexity and by the character of the costa subcosta, radius, and median veins which pass outwards towards the wing-apex in straight lines, and show no trace of divisions until well beyond the middle of the wing contrasting strongly

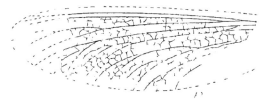

Fig 1 *Orthocosta splendens* bolton restoration of wing showing portion preserved and the character of the interstitial neuration, natural size —Middle Coal Measures (below the Top Hard Coal), Shipley Manor Claypit Ilkeston Derbyshire   Morton Collection Mus Pract Geol (nos 30,222 and 30,223)

with the areas occupied by the marginal divisions of the median cubital and anal veins   The inner two-thirds of the wing is marked by deep furrows in which lie the marginal branches of the median vein and the whole of the cubital and anal veins   The areas between any two veins in this region are markedly convex The differences in character of the outer and inner portions of the wing are emphasised by a line of fracture which fairly accurately separates the two   Its occurrence suggests that it has followed a natural line of weakness the more delicate inner part of the wing breaking away from the outer stronger portion

The costa subcosta, radius and median are all well-developed veins stout in structure and standing out in relief   The first three retain this evidence of strength over two-thirds of their length, the median vein showing signs of attenuation beyond the proximal third

The general structure of the wing indicates considerable powers of flight The outer margin, of which only a portion is preserved, appears as a stout slightly elevated and well-rounded ridge

The subcosta agrees in general character with the costal remnant, and is

straight to the wing-tip parallel with the outer margin, and not far removed from
it   Basally it appears to be united to the radius

The radius gives off the radial sector at about 10 mm from its base, and then
passes out to the wing-tip, at no point being more than 10 mm from the subcosta
The radial sector diverges widely from the radius, the two enclosing a long, narrow
triangular area   At 50 mm beyond its origin it divides into two equal branches,
which diverge to a distance of 3 mm and then become parallel   The direction of
the branches of the radial sector is such that they must have ended on the inner
half of the wing-apex, the outer branch probably forking again before the wing-
apex was reached   The median vein consists of two branches, the common origin
of which is not shown, owing to the absence of the base of the wing   To the
middle of the wing the two branches are parallel, and both pass beyond this
point before branching   The outer branch attains a length of 66 mm before
branching, and then gives off four branches on its outer side the last arising close
to the margin

The second of the four branches bifurcates   The inner branch of the median
gives off one forward branch only, which forks into a feeble twig dying out in the
polygonal meshwork, and into a stronger division which reaches the margin   The
two branches of the median are united at the base of the wing by an oblique
commissural vein which comes off at an acute angle from the outer branch, and
passes obliquely to the inner branch

The basal portion of the cubitus has been lost, so that its branching is not
readily determinable   The main vein seems to have divided near the base into
two equal branches, which curve down to the inner wing-margin, bifurcating
twice in each case before the margin is reached   The eight marginal twigs of the
cubitus thus produced enclose the middle third of the inner margin   The anal
area is wholly destroyed on the wing, and only a trace of one vein can be distinguished on the counterpart   This is a narrow deeply-sunk vein which gives off
feeble off-shoots on both sides   It diminishes in strength, so that the last portion
of its course can only be determined with difficulty   The anal area is comparatively
narrow and small, and exhibits few veins

Family PTERONIDIDÆ, Bolton (emend Cockerell)

1912   Bolton Quart Journ Geol Soc, vol lxviii p 314

Wings long and narrow   Outer margin arcuate   Median and cubitus with
divergent branches

The attempt to classify these wings is difficult   That they possess affinities
with the genus *Polycreama* is certain but they are more simple and more
Dictyoneurid in type   As Dr Handlirsch observes, it is not possible to refer them
to the family Polycreagridae, and they must be taken as the type of a new family

Prof. T. D. A. Cockerell has pointed out to me that the family name was originally wrongly written "Pteronidae," and that the generic name was antedated by *Pteronidea* Rohwer, 1911. He thinks that the difference is sufficient to justify retention, although it is only a single letter ('Proc. U. S. Nat. Mus,' vol. xlix, p. 469, 1915). Handlirsch ('Revision Palaozoischen Insekten' 1919, p. 15) has made the same correction, but seems unaware that Cockerell forestalled him.

Genus **PTERONIDIA**, Bolton

1912.   *Pteronidia* Bolton. Quart. Journ. Geol. Soc. vol. lxviii, p. 315.

Generic Characters.—Wings three times as long as wide. Radial sector with five simple, backwardly directed twigs. Median dividing low down into two principal branches, and ending on the inner margin in five twigs. Cubitus stout, divided into two principal branches the foremost with two twigs and the hinder ending in four. Interstitial neuration of thin oblique nervures.

Pteronidia plicatula, Bolton.   Plate I, fig. 3, Text-figure 2

1912.   *Pteronidia plicatula* Bolton. Quart. Journ. Geol. Soc. vol. lxviii, p. 313, pl. xxvi, figs. 1—3.

Type.—Incomplete wing and impression, Moysey Collection Museum of Practical Geology, Jermyn Street (nos. 30,224 and 30,225).

Fig. 2.—*Pteronidia plicatula* Bolton, restoration of left wing, natural size.—Middle Coal Measures below Top Hard Coal. Shipley Manor Claypit, Ilkeston, Derbyshire. Moysey Collection Mus. Pract. Geol. (nos. 30,224 and 30,225)

Horizon and Locality.—Middle Coal Measures (below the Top Hard Coal), Shipley Manor Claypit, Ilkeston, Derbyshire.

Specific Characters.—Wings triangular in outline, with subacute tip, strongly plicated. Radial sector dividing into five simple twigs. Median with two principal branches and five marginal twigs. Cubitus wide-spaced, two principal branches the inner doubly forked.

Description.—This is a long delicate wing, of which the outer marginal portion and the base are missing. It was probably about 70—75 mm. long, with a basal width of 20 mm., reduced to 8 mm. near the tip.

The outer margin must have been flatly convex, while the inner margin is nearly straight The tip of the wing is acute The costal and subcostal veins are not seen The radius is represented by a single forked vein not shown in the figures The radial sector is unbranched until well beyond the middle of the wing, beyond which it gives off four (possibly five) backward branches, which occupy the greater part of the tip of the wing and a portion of the distal inner margin

The median is represented by two branches the outer dividing into two twigs and the inner into three

The marginal area occupied by the median is a little longer than that of the radial sector, and the two extend over the distal half of the inner wing-margin

The cubitus is a relatively short and stout vein, dividing into two main branches, the foremost having two twigs, and the hinder dividing into four by a double bifurcation Owing to the wide divergence of its branches the cubital area is larger than that of the median The anal area was large, but the whole surface has been pitted by attempts to clear the matrix, and as a result the presence and character of the veins cannot be distinguished with certainty It is possible that certain surface-indications are evidence of two simple widely separated veins The wing-surface exhibits a strong plication

Viewed obliquely, the wing shows a series of ridges formed by the vein with V-shaped intervening sulci, which only flatten out close to the wing-margin Springing from the principal veins is a close series of fine cross-nervures, obliquely disposed in the direction of the wing-margin The portion of the wing preserved is sufficient to indicate that as a whole it was triangular in outline

Family HYPERMEGETHIDÆ, Handlirsch

1906 Handlirsch, Proc U S National Museum, vol xxix, p 672

Costa marginal, costal area broad , radius simple, radial sector present , the median probably dividing at the base into two or more main branches, the first of which may become united to the radius Cubitus forked near the base, with its branches widely spaced Anal veins few, and anal area not exceeding one-third of the inner margin

Handlirsch established this family for the inclusion of a gigantic wing of which only the basal half is known The total length of the whole wing was estimated to be 120 mm , the basal half having a length of 60 mm The discovery in the Coal Measures of Durham of a nearly similar wing adds to our knowledge of the group, and enables the family to be defined with more accuracy than was at first possible

The general assemblage of characters found in the Hypermegethidæ is I believe, highly suggestive of the Protodonata, but no definite conclusions can be formulated until the whole wing is known

Genus **HYPERMEGETHES**, Handlirsch

1906 *Hypermegethes* Handlirsch Proc U S National Museum vol xxix p 672

Generic Characters.—Costal border feebly convex, subcosta and radius close together along the greater part of their length   Cubitus composed of two possibly more parallel and simply forked branches, divided near the point of origin   Anal veins few   interstitial neuration of fine irregularly anastomosing nervures

Hypermegethes northumbriæ, Bolton   Plate I, fig 4, Text-figure 3

1917 *Hypermegethes northumbriæ* Bolton Quart Journ Geol Soc vol lxxii p 55 pl iv figs 2 and 3 and woodcut in text

Type.—Portion of the basal half of a left wing in counterpart   British Museum (no In 18524)

Fig 3 *Hypermegethes northumbriæ* Bolton, suggested outline of left wing restored showing portion preserved, slightly less than natural size—Coal Measures (shale above the Crow Coal Phœnix Brickworks Crawcrook Durham Brit Mus (no In 18524)

Horizon and Locality.—Coal Measures (shale above the Crow Coal), Phœnix Brickworks, Crawcrook, Durham

Specific Characters.—Costal area wide, and crossed by an irregular meshwork of small veins   Median united with the radial sector basally, and giving off an inner branch which unites with the cubitus   Cubitus with a short stem forking into two equal and widely separated branches   Anal veins simple, and widely spaced

Description.—The fragment is little more than a third of the whole wing, and its characters can only be determined with difficulty   The two halves of the nodule do not coincide, and the outline drawing of the wing is built up from details supplied by both halves   The inner margin anal area and base of the wing have been lost, so that a little less than two-thirds of the outer portion of the basal half of the wing is present

The portion of wing remaining being 63 mm long with a depth of 34 mm at its widest part the whole wing must have had a length of about 126 mm, or

5 m , and the whole insect a span of wing of nearly 11 m   The wing-fragment shows the basal portions of the costa, subcosta, radius, radial sector, median, and cubitus and possibly a trace of an anal vein

Much of the finer detail of the wing is not seen until the fossil is immersed in water—a mode of treatment suggested to me by Dr F A Bather, F R S , who had previously photographed the wing in this manner and brought out details of which by ordinary methods, I could not find a trace

The costa is moderately convex from its base to a distance of about 30 mm , beyond which it becomes straight   Separated from the outer or costal border by a wide area basally is the subcosta, an extremely feeble and hardly distinguishable vein   It passes straight to the outer margin some distance beyond the middle of the wing   The radius arises close to the subcosta and is parallel with it   It gives off two branches posteriorly, the proximal branch passing obliquely towards the inner side of the wing-apex, while the second or distal branch arises from the radius a little further out, and keeps parallel with it

I had formerly considered the proximal branch of the radius to be the radial sector, and the distal a branch of the radius , but as the interstitial neuration now shows the specimen to be closely related to *Hypermegethes schucherti*, the proximal branching vein must be regarded as the main stem of the median which has entered into union with the radius, and the distal branch as the radial sector

Regarding the proximal offshoot of the radius as the median vein, it diverges widely from the radius, giving off a forward twig parallel with the radial sector, and then continues inwards and unites with the next vein, separating again a little further out on the inner side   A comparison of this assumed median vein with that of *H schucherti*, Handl , is instructive   In the latter species the branching of the main stem of the median arises nearer the base of the wing than does the branching off of the radial sector.   It is therefore somewhat in the position of the starting-point of the median vein as a free structure in the present wing   The median vein of *H schucherti* Handl , has however, no union with the radius, the main stem running out towards the wing-apex parallel with the inner branch of the radius giving off a backward branch of the cubitus, which passes obliquely inwards and unites with the anterior branch   In the middle of its length it gives origin to a twig running parallel with the main stem, and midway between that vein and the cubitus.   The course of the median vein in the specimen here described is exactly similar to that of the branch of the median vein in *H schucherti*, Handl , except that the inner branch not only unites with the cubitus, but crosses it

The condition may be summarised by saying that, in *H schucherti* Handl , the median vein is entirely free and gives off an inner branch dividing into two twigs, of which the inner unites with the cubitus   In the specimen here described the median is united with the radius for some little distance, becoming free before the

5

radial sector is reached, and then forking, the inner twig uniting with and crossing the cubitus

The cubitus has a short stout basal stem forking into two equal and somewhat widely separated branches the outer uniting with the inner branch of the median The inner branch is forked just before the broken edge of the wing is reached Nearer the inner margin of the wing are traces of two other veins The first has its basal portion missing and follows a course parallel with the inner branch of the cubitus It is strongly forked The remaining vein is represented by three detached fragments If Handlirsch's interpretation of the wing of *H schucherti* is followed, we should regard both these veins as anal I am, however, of opinion that, while the innermost fragmentary vein may be anal, the forked vein by its position its forked character and stoutness must be regarded as a portion of the cubitus I am likewise of opinion that the first, and possibly the second, of the veins marked as anal in Handlirsch's figure of *H schucherti*, ought also to be classed as cubital Both in *H schucherti*, and in this specimen, the anal area would have an enormous development and occupy most of the wing-margin, if the veins alluded to were wholly anal in character I feel sure that if the vein nearest the cubital had been better preserved, it would be found branching off from the cubitus

The interstitial neuration is typically that of *Hypermegethes* The intercostal area is filled with an irregular meshwork of fine nervures, with a tendency on the outer and inner sides to a transverse arrangement Between the median and the cubital veins the interstitial neuration consists of short straight and transverse nervures, and feeble traces of similar nervures can be seen in the median area The cubital area is filled with a meshed neuration larger and more regular than that of the intercostal area, and this seems to continue into the anal area

Family CRYPTOVENIIDÆ, Bolton.

1912   Bolton Quart Journ Geol Soc, vol lxvin, p 315

Wings short and broad Apex rounded, costal vein marginal, subcosta feeble, and extending to near apex Radius simple, radial sector and median with few divisions, cubitus with two main branches

Among the insect-remains discovered by the late Dr L Moysey at the Shipley Manor Claypit is a small wing, 16 mm long, unlike any previously known It is typically Palæodictyopterous and agrees remarkably well with Dr Handlirsch's type-figure (Mitth Geol Gesell Wien, vol in 1910 p 505, fig 1) It differs from that form in the greater division of the cubitus, which ends in five twigs instead of three The greatest depth of the wing was also in all probability nearer the base than in his figure With the genus *Ithymodictya*, Handlirsch ('Amer Journ Sci,'

[4], vol xxxi, 1911, p 298), the relationship is even closer, as in that genus the costa and subcosta are close together, the radial sector arises low down and is divergent from the radius the median is a simple vein ending in three branches, while the cubitus is almost identical in its divisions, the difference being that the first forking arises at a higher point than that in *Athymodictya parva* Handl and that the inner simple ramus of the outer branch comes off a little below the middle of the wing The anal veins number four in *A parva* as against five in the wing under consideration The wings are almost equal in size It is not possible, however, to refer this wing to *Athymodictya* owing to the character of the interstitial neuration In *Athymodictya* this is reticulated, while in the British wing the interstitial neuration is apparently made up of transverse nervures If this point can be clearly determined, the wing may have affinities with the Homoiopteridæ or with the Lithomantidæ, but it cannot be referred to either of these families Neither Dr Handlirsch nor I could satisfactorily refer the wing to any known family, and I therefore founded the family Cryptovenidæ to receive it

Genus **CRYPTOVENIA**, Bolton

1912 *Cryptovenia* Bolton Quart Journ Geol Soc, vol lxviii, p 315

Generic Characters —Wings twice as long as wide Median, cubitus and anal veins curving sharply inwards Interstitial neuration of straight cross-nervures

Cryptovenia moyseyi, Bolton Plate II, fig 1, Text-figure 1

1912 *Cryptovenia moyseyi*, Bolton, *loc cit* p 315, pl xxxii figs 1—6

Type —Incomplete wing, in counterpart Moysey Collection Museum of Practical Geology Jermyn Street (nos 30,226 and 30,227)

Horizon and Locality —Middle Coal Measures (below the top Hard Coal) Shipley Manor Claypit, Ilkeston, Derbyshire

Specific Characters —Costa feebly curved distally, subcosta reaching outer margin near wing-apex Radius close to and parallel with the subcosta, radial sector arising from radius in basal half of the wing, the five branches occupying the inner half of the wing-tip Median with the first branch undivided, and the second forking twice Cubital vein strongly arcuate two-branched, and ending in five twigs Anal veins five or more in number, and curving sharply inwards Wing pleated, and the interstitial neuration obscured by a mass of wrinkles

Description —A small delicate wing incomplete at the base which must have been very narrow and with the main veins crowded together The total length of wing preserved is 16 mm, and its maximum diameter, in the cubito-anal region, is 8 mm

The costal vein is stout and raised above the general level of the wing in the outer third of its length, where it curves backwards into the wing-tip. The subcostal is a weaker vein, parallel with and very close to the costa, dying out or joining the outer margin at the apex. The radius is a powerful undivided vein, parallel with the subcosta and reaching the middle of the wing-apex. The radial sector is given off about the middle of the length of the wing and forks into two equal branches which again fork before reaching the margin; the outermost branch of the second forks again and divides just before the wing-apex is reached. The

Fig 4—*Cryptovenia moyseyi*, Bolton restoration of left wing, reconstructed from the wing fragment and counter-impression enlarged two and a-half times—Middle Coal Measures (below the Top Hard Coal), Shipley Manor Clypit near Derbyshire Moysey Collection Mus Pract Geol (nos 30,226 and 30,227)

inner half of the wing-apex is occupied by the five branches of the radial sector. The median vein is a comparatively simple structure forking low down below the middle of the wing into two nearly equal branches. The outer branch remains undivided, and gently curves to the inner margin. The inner branch divides twice, first at a point near the middle of the wing, and again before the margin is reached. The median vein therefore ends on the margin in four twigs, three members of which arise from the inner of the two main branches.

The cubital vein is strongly arcuate, dividing near the base into two branches, the outer forking twice and the inner once. The cubital vein therefore ends in five twigs.

Family MECYNOPTERIDÆ, Handlirsch

1906 Handlirsch, Die Fossilen Insekten, p 82

Handlirsch established this family to receive a large wing from the Middle Upper Carboniferous of Belgium, and placed it between the families Hypermegethidæ and Lithomantidæ. The characters of the family are based on those of the type-species, *Mecynoptera splendida*, Handl

Genus **MECYNOPTERA**, Handlirsch

1904 *Mecynoptera*, Handlirsch, Mem Mus Roy Hist Nat Belg vol iii p 7
1906 *Mecynoptera*, Handlirsch, Die Fossilen Insekten, p 82

General Characters —Wing three to four times as long as broad, veins of costal region specially compact, and thickened basally Costa, subcosta and radius closely approximated and tuberculated Radial sector and median well developed, and occupying a considerable portion of the wing-surface Cubitus small Interstitial neuration of transverse nervures in junction areas of principal veins, and with a meshwork further out

Mecynoptera tuberculata, sp nov Plate II, fig 2, Text-figures 5 and 6

1911 *Stenodictya lobata*, Baldwin (errore), Geol Mag [5], vol viii, p 75

Type —Portions of two fore-wings, and the cubito-anal portion of a hind-wing contained in a nodule of ironstone, 3 in long, and 1½ in wide, British Museum (no In 18,576)

Horizon and Locality —Middle Coal Measures (grey-blue shales at 135—180 feet above the Royley or Arley Mine), Spaith Bottoms, Rochdale, Lancashire

Specific Characters —Principal veins thickened at base and finely tuberculated Costal margin almost straight, much thickened at base, and covered with a fine

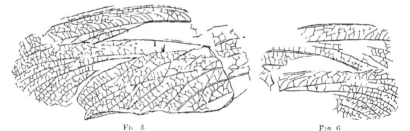

Fig 5 —*Mecynoptera tuberculata* sp nov diagram of remains of the two fore wings, and culato anal portion of a hind wing (the sub costa appearing as a fine line close to the radius and sending off three branches to the costal margin) enlarged one and a half times—Middle Coal Measures (above the Royley or Arley Mine), Spaith Bottoms, Rochdale, Lancashire Brit Mus (no In 18 576)

Fig 6 — *Mecynoptera tuberculata* sp nov diagram of fragmentary impression of the two fore wings enlarged one and a half times —Middle Coal Measures (above the Royley or Arley Mine) Spaith Bottoms, Rochdale, Lancashire Brit Mus (no In 18,576)

tuberculation Subcosta sunken, close to costa, and extending to the wing-apex Radius close to costa tuberculated, and elevated Radial sector arising low down, much branched. Median with two main branches, the first simple, the second with three twigs Cubital vein small

Description —The type-specimen comprises remnants of the two fore-wings, 60 mm long and 18 mm wide, and a portion of the cubital-anal area of a single hind-wing in one ironstone nodule A similar ironstone nodule from the same horizon and locality shows the radius and radial sector areas of the two wings of a second insect

When complete the wings must have had a length of 65—70 mm   They were originally identified with those of *Stenodictya lobata*, Brongniart, and are quoted as such in the faunal lists of the Sparth Bottoms deposits, but with no detailed description or figures

When the specimens were first examined they presented an apparently anomalous union or suppression of certain of the principal veins   Careful development has since shown that what was formerly considered to be the outer or costal margin of the wings was really the thickened and tuberculated radius vein, and that the outer margin, with the costa and subcosta, had been hidden under the matrix. A little of the outer marginal costa and the subcosta are now uncovered, and the wing-structure is therefore proved to be of a normal type

The three wings in the larger, more nearly complete specimen are superposed and fragmentary   The uppermost fore-wing is represented by two portions, one having a little of the outer or costal margin, the subcosta, and the greater part of the radius and radial sector   The second part consists of the median, and a portion of the cubitus   The second fragment is displaced backwards, allowing the radius and radial sector of the second fore-wing to be seen   The median-cubital area of the hind-wing is very thin and closely pressed on the rest, but the course of the veins is clearly discernible, and by their sharp inward turn indicates that the hind-wings were much broader than the fore-wings

Although the wings are thus broken up, displaced, and superposed, the parts missing in the one wing are present in the other, and it is possible to reconstruct their general character   The outer margin is almost straight, gently rounded into the base, and into the wing-apex   The subcosta is weak, and lies in a deep sulcus extending into the apex of the wing   The radius is a little elevated, thickened tuberculated in the basal third and in close proximity to the subcostal vein   The radius is a strong vein, raised above the level of the rest of the outer margin of the wing so that when the latter was hidden, it naturally appeared to be the outer marginal costal vein   It is much thickened in its basal third, tuberculated, and remains parallel with the subcosta in the apex of the wing

The wing-space taken up by these three veins is very small, and their close approximation and the coriaceous thickening of the costa and radius serve to give a considerable degree of strength and rigidity to the outer margin of the wing   The radial sector passes straight outwards to the apex of the wing, giving off three inner branches in one wing and two in the other the first branch only forking   The first branch soon forks, the outer fork dividing into two twigs, and the inner into three upon the distal inner half of the wing-apex   The median is a large vein, occupying the greater part of the inner half of the wing   It divides at its point of origin into an outer undivided branch which traverses the middle of the wing, and an inner branch which is widely separated from its fellow, and gives off three twigs to the wing-margin, the first forking in the middle of its length   Cubital veins are represented by two undivided elements which are directed obliquely to the margin. The inner margin of the wing is more curved than the outer, and merges into the apex. The interstitial neuration is remarkable. Between the costa, subcosta and radius it consists of short stout nervures crossing the areas somewhat obliquely, and in some cases arranged in **V**-shape. The area between the radius and radial sector is crossed by a numerous **S**-shaped series of nervures, which are joined up into a meshwork in the outer or radial half of the area. The area itself is very wide, and only equalled by that separating the two branches of the median.

The radial sector and the median occupy the greater part of the wing-surface.

The remaining areas are crossed by transverse nervures near the junction of branches with the principal veins, and further out by nervures which are joined up by zig-zag longitudinal branches which occasionally enclose polygonal cells. The interstitial neuration is very well developed, and must have added materially to the strength of the wings.

Affinities.—The close approximation of the costa, subcosta and radius and the coriaceous thickening and tuberculations of the principal veins, are characters which may rank as of generic importance. Less distinctive, but also characteristic, is the interstitial neuration of the transverse nervures and meshwork, with its minute inosulation.

The characters of the genus *Mecynoptera* are in closest agreement with those of this insect. There is the same approximation of the costa, subcosta and radius, the radial sector is widely spaced from the latter, and the interstitial neuration consists of transverse nervures near the junctions of the principal veins and their branches, and of a meshwork further out. The general outline of the wing is also the same.

M. splendida, Handlirsch, throws light on one point which had proved a difficulty in the determination of the wings. It has the radial sector of large size and much branched, the divisions occupying all the inner half of the apex of the wing, and extending out on the inner margin into the area usually occupied by the outer branches of the median. Prior to noting this feature in *M. splendida*, I had formed the opinion that the first division of the radial sector was the median, which had united with the radial sector and the radius.

The broken condition of the Rochdale wings prevents absolute determination of this point, but by analogy I conclude that the whole vein which branches off from the radius basally is the radial sector, and the next vein the median. This conclusion also removes the insect from the neighbourhood of *Eucynatodes danielsi*, Handlirsch, a wing discovered in the Coal Measures of Mazon Creek, near Morris, Illinois, U.S.A., which possesses the same character of interstitial neuration, but with the divisions of the median vein stretching out to the wing-apex, and the costa, subcosta and radius more widely separated.

Unless it can be shown by the discovery of a whole wing that the reconstruction now attempted is faulty, the balance of evidence is in favour of the provisional reference of the Sparth Bottoms wing to the genus *Mecquoptera*

Family INCERTÆ SEDIS

Genus **PALÆOMANTIS**, Bolton

1917   *Palæomantis* Bolton Quart Journ Geol Soc, vol lxxi p 52

Generic Characters —Wings short, twice as long as broad   Apex well rounded   Radius, radial sector and median all powerful veins with few divisions.   Median with two main branches   Anal veins directed almost straight inwards   Interstitial neuration forming an irregular meshwork

Palæomantis macroptera, Bolton   Plate II, fig 3, Text-figures 7 and 8

1871   " Wing of large insect,' Higgins Pres Add Liverpool Naturalists' Field Club, vol ii p 18
1917   *Palæomantis macroptera* Bolton Quart Journ Geol Soc vol lxxi, p 48 pl iii, figs 3–4, text-figs 2—3

Type —Remains of two wings in nodule, Liverpool Museum

Horizon and Locality —Middle Coal Measures, Ravenhead railway cutting, near St Helens, Lancashire

Specific Characters —Divisions of radial sector occupying almost the whole tip of the wing   Divisions of median and cubital veins occupying distal two-thirds of inner margin

Description —The larger half of the nodule has lost a portion which contained the tip of the right wing   The proximal third of the left wing remains, with its dorsal surface uppermost and its ventral surface closely applied to that of the right wing   It is evident that the whole of the two wings was contained in the nodule, but, when the latter was split open, a thin film of ironstone carried away the middle and distal portions of the left wing

One unusual feature in the position of the wings is that they lie with their ventral surfaces apposed   To bring them into this position one must have become bent under the body, instead of falling sideways across the thorax of the insect   The body of the insect would thus, if the wings still remained attached, lie between the two   No trace of the body can be seen, but the left wing has a deep inward flexure, such as it would naturally acquire if the body had been carried round with the right wing and pushed into the anal area of the left wing   Had the body been carried round in this way the right wing would not coincide in position with the left, but be thrust further out, which is actually the case, the outward displacement ot the right wing as compared with that of the left, being at least 20 mm. The wings are remarkably wide—a feature which we usually associate with hind-wings.

The outer third of each wing is supported by strong and fairly rigid veins, becoming more slender as they pass towards the wing-tip. The inner margins of the wings are more membranous, and were evidently lacking in a rigidity equal to that of the outer margins.

A little more than the proximal third of the *left wing* is present and in good preservation. It is 42 mm. long and about the same in width. The impression on the other half of the nodule shows apparently all but the wing-apex. The main stems of the costa, subcosta radius and median are all stout, and in relief on the wing-surface. The cubital and anal veins are but half the thickness of the former and lie in shallow grooves. The costal vein forms the outer margin of the wing, which is slightly convex forwards, and slopes gradually into the apex. The subcosta is

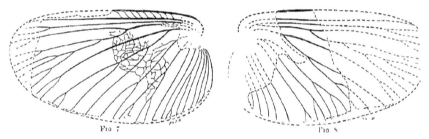

Fig 7   Fig 8

Fig 7.—*Palæomantis macroptera*, Bolton, restoration of left wing, with restored outline, natural size. Middle Coal Measures. Ravenhead Railway Cutting near St Helens, Lancashire. Liverpool Museum

Fig 8.—*Palæomantis macroptera*, Bolton, restoration of right wing, natural size.—Middle Coal Measures, Ravenhead Railway Cutting, near St Helens, Lancashire. Liverpool Museum

separated from the outer margin by a costal area 1 mm. broad. It passes almost in a straight line to the outer edge of the wing-apex, and is joined to the costa by a series of oblique transverse branches nearly parallel with each other. The radius arises close to the base of the subcosta, and diverges a little from it in its course to the apex, forking once just before the broken edge of the wing is reached. The main stem of the radial sector probably arises from the radius close to the base of the latter, but the exact point cannot be determined, owing to the base of the wing being broken away. The radial sector forks three times, and each of the resultant twigs again divides on the wing-apex. The median vein which is stout basally, divides very low down into two branches the outer of which has a feeble forking near the inner margin while the inner branch forks three times. The cubitus is a much weaker vein than its fellows, and divides near its point of origin into two branches. The first branch forks once and the second twice. Seven anal veins are present, all undivided except the third, which forks twice. The interstitial neuration is in the form of a meshwork except in the intercostal area, where are the transverse cross-nervures already noted.

Much less of the *right wing* remains than of its fellow, but sufficient is present to show that the neuration was not quite the same. The median ends in five divisions in place of six, and the cubitus has seven final branches instead of six. The greatest width of the wing is along a line drawn from the outer margin to the middle of the cubital area on the inner margin, the width being 10 mm. The absence of the apex in each wing is unfortunate, as it renders the outline of the whole wing a little uncertain. The shape of the nodule indicates that very little of the wings is missing if they were wholly included in the nodule as seems probable. The somewhat semicircular inner wing-margin and the short wide wings indicate a broadly rounded wing-apex.

Affinities.—The interstitial neuration is much like that of *Hypermegethes*. The great width of the wings is a character usually associated with the hind-wings of members of the family Lithomantidæ although in this case the wings lack the distal attenuation noticed in the hind-wings of that family. With *Titanodictya jucunda* Scudder, there is a close relationship both in the general character of the main veins, the interstitial meshwork in all areas other than intercostal and subcosta-radial, and in the presence of the same oblique cross-nervures in the intercostal area in these wings.

The true systematic position of the genus seems to lie between the Lithomantidæ and the genus *Titanodictya*, and closest to the latter. Because of the greater development of a meshwork neuration between the main veins, and the limitation of cross-nervures to the intercostal area I regard this genus as more primitive than any of the Lithomantidæ, but closely related thereto.

Family LITHOMANTIDÆ, Handlirsch

1906   Handlirsch Proc U S National Mus vol xxix, p 673 also Die Fossilen Insekten, p 82

This family is closely allied, by wing-structure, to the Dictyoneuridæ. The branching of the main veins has proceeded further than in the Dictyoneuridæ, and the body, where it has been preserved shows striking differences. The family is represented in the Coal Measures of Great Britain and North America and in the Upper Carboniferous of continental Europe.

Genus **LITHOMANTIS**, Woodward

1876   *Lithomantis* H Woodward Quart Journ Geol Soc vol xxxii, p 60

Generic Characters.—Large insects with two pairs of flying wings, the hinder pair double the width of the anterior pair. Prothorax produced into a central rostral-like process, acutely pointed, and expanded into two lateral bulbous lobes. Thorax wide. The interstitial neuration consists usually of stout transverse nervures, occasionally forking or in a loose meshwork.

Lithomantis carbonarius. Woodward. Plate II. fig. 4; Text-figure 9.

1876. *Lithomantis carbonarius*, Woodward, *loc. cit.*, p. 60, pl. ix, fig. 1.
1893. *Lithomantis carbonarius*, Brongniart, Faune Entom. Terr. Prim., p. 489, fig.
1906. *Lithomantis carbonarius*, Woodward, Geol. Mag. 5, vol. iii, p. 25, fig. 3 (*non* fig. 1).
1906. *Lithomantis carbonarius*, Handlirsch, Die Fossilen Insekten, p. 83.

Type. —Portions of the fore- and hind-wings, with prothorax, and a left anterior leg in nodule; British Museum (no. I. 8118).

Horizon and Locality. —Coal Measures; Scotland.

Specific Characters. —Hind pair of wings double the width of the fore pair. Outer wing-margin straight, costa and subcosta close together; radius a weak vein, giving off the radial sector far out. Median small. Cubitus a powerful vein with numerous widely spaced branches. Anal veins six or more.

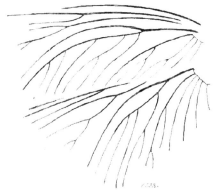

Fig. 9.—*Lithomantis carbonarius*, Woodward; diagram of neuration of the left fore- and hind-wings, natural size.—Coal Measures, Scotland. Brit. Mus. (no. I. 8118).

Description. —The type-specimen lies on the surface of one half of an ironstone nodule, the counterpart being lost. It was obtained by Mr. Edward Charlesworth from the Coal Measures of Scotland, but the locality and horizon are not known.

The remains consist of portions of two pairs of wings, those of the left side being most nearly complete. In no case is the apex of the wing or the inner margin preserved. Lying in front of the wings are a pair of very large convex lobes which Dr. Woodward regards as part of the prothorax, and in front of these is a roughly quadrangular structure prolonged forward in the middle line into a styliform process or rostrum. Woodward describes the latter as " the small head with its eyes,' but no definite trace of the latter are observable. The ' head ' is small, not more than 7 mm. wide and between 5—6 mm. long at the sides; medially it is prolonged into the rostral process which is about 8—9 mm. long and ending in a sharp point. I am unable to distinguish any dividing line between the ' head ' and the prothorax; on the other hand the margin of the latter is continuous with and inseparable from it. The marking which Woodward has regarded as representing eyes is, I believe, the thickened margin. I am of opinion that the whole structure is wholly prothorax, and that the head lies concealed beneath. The main mass of the prothorax is 30 mm. wide with a flattened margin, best seen on the left side. Within the flat margins rise two low dome-shaped lobes separated by a wide hollow in front, their margins meeting in an obtuse angle posteriorly. The median edge of each lobe dips sharply into a wide median hollow, and from each of these edges arises a series of veins which spread out to the lateral margins of the lobes. The areas between the veins are occupied by a fine meshwork of smaller veins.

A trace of the mesothorax is shown between the bases of the fore-wings as a slight transverse bar, a small rounded tubercle lying in front and a little to the left of the middle line.

The left fore- and hind-wings are the most nearly perfect, the hind-wing being 56 mm. long with a greatest width of 30 mm. The fore-wing is a little shorter and narrower. The outer margin appears to have been straight and the costa and subcosta closely approximated. Traces of both veins are present. The radius is a straight thin vein not far removed from the subcosta along its whole length and giving off the radial sector beyond the middle of the wing. The radial sector comes off at an acute angle, going out to the wing-apex and keeping closely parallel with the radius. The median is somewhat inconspicuous owing to the great length of the main stem and the narrow areas which bound it between the radius and the prominent cubitus. It gives off two outer branches before reaching the broken edge of the wing. The median vein of the hind-wing has three outer branches, the first arising nearer the base than the point of origin of the radial sector, whereas in the fore-wing the first branch arises distally to the origin of the radial sector. The median of the hind-wing is a more important vein than its fellow in the fore-wing and occupies a greater area owing to its greater inward curvature. The cubitus is a powerful vein with widely spaced divisions, the first branch, both in fore- and hind-wings, coming off from the main stem on the outer side and low down near the base of the wing, and then passing in a bold convex sweep down to the distal portion of the inner margin. On its inner side the cubitus gives off five branches, the fourth forking in the middle of its length. The branches arise at irregular distances, and the main stem reaches the margin far out towards the wing-apex. The cubitus of the hind-wing gives off a large outer branch which is almost equal in strength to the main stem. This arises even nearer the base than its counterpart in the fore-wing, but its great length is masked by the development of the median. Beyond the origin of the outer branch the cubitus gives off three inward branches, the first forking. Owing to the strong inward curvature of the cubitus and its branches, the succeeding anal veins are directed almost straight inwards. The anal veins are six in number, the first forking twice, and ending on the margin in three branches.

The remaining veins are undivided. The cubital and anal veins of the fore-wing are more obliquely disposed than those of the hind-wing, and the latter has an anal area much larger than that of the fore-wing.

The interstitial neuration consists of short stout cross-nervures, which occasionally fork, and in the wider areas unite to form a meshwork.

A triangular area is marked off from the base of each wing by a deep furrow. This Woodward correlates with a similar area in the wing of *Gryllacris* (*Cordulis*) *brongniarti*, which Swinton (Geol. Mag., [2] vol. i, 1874, p. 337, pl. xiv, fig. 3) has described as a stridulating organ. As stated elsewhere I am of opinion that Swinton's conclusions in the case of *G. brongniarti* were founded on a misinterpretation of the wing, and that no stridulating organ, or similar structure, is present. In the case of *L. carbonarius* such a structure would apparently be useless, as none of the wings could come into such close apposition as would allow the structure to be used. I am unable to determine its purpose or significance, unless it be a portion of the musculature attachment of the wing.

Traces of a fore-leg are present, projecting from beneath the left lobe of the prothorax. Its structure is too indefinite for description.

The wings were probably one-fourth or one-third longer than the portion preserved—an estimate which would make each complete wing about 70 mm. long with a spread of about 140 mm., or 6 inches.

Genus **LITHOSIALIS**, Scudder

1881   *Lithosialis*, Scudder Geol. Mag. [2] vol. viii p. 299

Generic Characters.—Wings three to four times as long as wide. Outer and inner margins almost parallel, intercostal area wide basally, and diminishing to extinction at apex of wing. Radius simple, median with two main branches, and cubitus large, anal veins few and oblique.

The wing on which this genus is founded is interesting as being the first discovered in British Palæozoic rocks, except the problematic examples mentioned by Lhuyd and not otherwise known. Supposed to be a plant, the wing was first sent by Mantell to Brongniart, who in turn referred it to Audouin. The latter recognised its insect character, and brought it before the Entomological Society of France, the Academy of Sciences, and the Assembly of German

Naturalists at Bonn   Audouin described the wing as that of an unknown
Neuropterous insect allied to *Hemerobius*, *Semblis* and especially to *Corydalis*
Mantell (= Audouin) described it as closely resembling a species of living *Corydalis*
of Carolina   Swinton states that Mantell purchased the fossil at a sale of
Parkinson's collection although Mantell (Medals of Creation, p 554) says that
he discovered" it in a nodule from Coalbrookdale   Possibly the nodule had
previously formed part of the Parkinson Collection   Both the figures by Mantell
and Murchison are badly drawn and it was not until 1874 that a reliable drawing
was published by Swinton   The latter author devoted considerable attention to
a " serrated vein" at the base of the wing which he regarded as a stridulating
organ   He therefore referred the wing to the Orthoptera, and to the genus
Gryllacris   Scudder threw considerable doubt on Swinton's conclusions and
showed that such an organ so placed could not have been of any service   As
we shall see later, the supposed stridulating organ is merely a torn edge of the
base of the wing   A study of the wing-structure convinced Scudder that the
wing was most closely related to the *Lithomantis carbonarius* Woodward   Being
generically distinct from *Lithomantis*, and from living types, he gave the insect
the generic name of *Lithosialis*

Lithosialis brongniarti (Mantell)   Plate III fig 1 , Text-figure 10

1833   *Corydalis ?* Audouin Ann Soc Ent France vol ii, Bull p 7
1836   *Corydalis ?* Audouin Buckland, Bridgewater Treatise vol ii p 77
1844   *Corydalis brongniarti* Mantell, Medals of Creation ed 1 vol ii, p 578 figu 124, fig 2
1854   ' Sialide , Pictet Traité de Paléontologie ed 2, p 377 pl xl fig 1
1867   *Corydalis*, allied to Murchison Siluria ed 4 p 300 woodcut 80
1871   *Corydalis brongniarti* Woodward Geol Mag vol viii p 387 (name only)
1874   *Corydalis brongniarti* Swinton, Geol Mag [2 , vol i p 339 pl xiv, fig 3
1875   *Gryllacris (Corydalis) brongniarti* Woodward, Geol Mag [2] vol ii, p 622
1876   *Corydalis brongniarti*, Woodward Quart Journ Geol Soc , vol xxxii p 62
1880   *Corydalis brongniarti* Nowak, Inhib k k Geol Reichsanst Wien, vol xxx p 73 pl ii fig 4
1881   *Lithosialis brongniarti* Scudder, Geol Mag 2 vol viii p 290
1885   *Lithosialis brongniarti*, Scudder, Mem Bost Soc Nat Hist vol iii p 220 pl xvii figs 1 2, 8 9
1887   *Protogryllacris brongniarti* Brongniart, Bull Soc Amis Sci Nat Rouen 3 ann xxi p 50
1893   *Lithomantis brongniarti* Brongniart Faun Entom Terr Prim p 371 figs 17 18
1906   *Lithosialis brongniarti* Handlirsch Die Fossilen Insekten p 84 pl v fig 13

Type —A left fore-wing , British Museum (Mantell Coll , *olim* Parkinson Coll ,
no 11,619)

Horizon and Locality—Coal Measures , near Coalbrookdale Shropshire

Specific Characters —Radial sector with not more than three branches   median
much branched and occupying much of the distal inner wing-margin   Cubitus
with three divisions   Anal veins few, and very oblique

Description—The left fore-wing measures 61 mm long and 18 mm wide across the distal third. The apex, the inner margin and a very small portion of the base are missing. The wing appears to have been strap-shaped with a blunt apex, the inner margin curving most into the apex.

The outer margin is feeble, almost straight over the greater part of its length and slightly sloped backwards. Basally it dips abruptly inwards to the point of attachment. The subcosta is widely spaced from the outer or costal margin proximally, the two gradually approaching each other until they meet at the wing-apex. The radius passes straight out to the apex, giving off the radial sector before the middle of the wing is reached; the radial sector remains undivided up to the distal third of the wing, beyond which it forks twice. The median vein divides a little further out than the radial sector, forming two strongly divergent branches, an outer large branch with two large forward twigs, and a small inner branch which forks, the inner twig forking again. The median, therefore, ends on the inner margin of the wing in six twigs.

Fig. 10.—*Lithosialis brongniarti* (Mantell). diagram of neuration of left fore-wing, natural size.— Coal Measures, near Coalbrookdale, Shropshire. Mantell Collection *olim* Parkinson Collection Brit. Mus. (no. 11 619)

The cubitus divides into three branches, the first arising low down, and passing in a bold sweep beyond the middle of the inner margin. The anal veins are five in number, all but the first being undivided. The first anal forks near the base, the outer branch forking again.

The whole wing is covered by a numerous series of strong transverse nervures, rarely branching, and usually crossing at right-angles to the main veins. A few are oblique or curved.

The "file" or 'serrated vein," described at length by Swinton, and by him correlated with that present in recent forms of *Gryllacris*, appears to be nothing more than an irregular torn edge of the basal part of the wing, which is also lifted up a little above the general level. The torn edge extends along the line of the median vein for a short distance and wholly lacks the symmetry and detail given to it by Swinton.

Affinities—Although closely allied to *Lithomantis carbonarius* Woodw., this wing differs in the great width basally of the intercostal area and in the straighter course of the cubitus and anal veins, due to the greater length of the wing. The interstitial neuration is much the same.

Genus **PRUVOSTIA**, novum

Generic Characters.—Fore-wings two and a half times as long as wide, outer and inner wing-margins almost parallel, wing-apex well rounded. Subcosta widely removed from the costal margin basally, and apparently not connected with the costa or radius. Radius straight with divergent radial sector. Radial sector, median and cubitus all distally branched. Anal veins few. Interstitial neuration of straight cross-nervures.

Certain features of this wing are suggestive of the Protorthoptera. These are the wide basal intercostal area, the basal origin of the radial sector, and the remote branching of the radial sector, median and cubital veins.

It is, however, more nearly allied to *Lithosialis* than to *Metria audis*, Handl, for example, among the Protorthopteroids. The great length before division of the main stems of the radial sector, median and cubitus, then strong divergence, and the many branches of the median, form an assemblage of characters not elsewhere known, and certainly deserving of generic recognition.

It is with pleasure that I attach to this genus the name of Dr. P. Pruvost, of Lille University, in recognition of his valuable work on the fossil insects of the north of France.

Pruvostia spectabilis, sp. nov.   Plate III, fig. 2.   Text-figure 11.

Type.—A left fore-wing, British Museum (Johnson Collection no. 1,15,891).

Horizon and Locality.—Middle Coal Measures (clay ironstone nodules in the Binds between the 'Brooch' and 'Thick' coals), Coseley, near Dudley, Staffordshire.

Fig. 11.—*Pruvostia spectabilis*, sp. nov., diagram of neuration of left fore-wing, natural size.—Coal Measures (clay ironstone nodules in the binds between the 'Brooch' and 'Thick' coals), Coseley, near Dudley, Staffordshire.   Johnson Collection, Brit. Mus. (no. I, 15,891).

Specific Characters.—Outer margin convex at base and inclining inwards to the apex. Subcosta straight, not reaching the apex of wing. Radius straight, giving off radial sector near base, radial sector twice branched, median divergent three times branched, the first two branches forking. Cubitus dividing basally into two, the outer forking three times, the inner branch once.

Description.—A left fore-wing 52 mm. long and 24 mm. wide, almost covering the surface of one half of the median plane of a reddish-brown ironstone nodule. A little of the base of the wing is missing, while a part of the wing-apex is concealed

by a film of ironstone which cannot be removed   Fortunately the slight film of
ironstone does not hide the outline of the wing, and it is evident that the outer and
inner wing-margins are almost parallel while the apex is well rounded   The outer
margin merges by a well-rounded contour into the wing-apex   The subcosta is
well-marked widely removed from the outer border proximally, and lies in a shallow
groove formed of the intercostal and radial subcostal areas   These two areas are
reduced distally to less than half their proximal diameter owing to the backward
inclination of the outer margin   The subcosta seems to die out near the radius
about 7 mm from the wing-apex and gives off a numerous series of irregularly
spaced and forwardly directed nervures, which are at first straight and then curved
towards the wing-apex   At their point of origin the cross-nervures are very
distinct, but they thin and occasionally die out before reaching the outer margin
The radius is a strong vein, passing perfectly straight out from the wing-base to
the outer part of the wing-apex   The radial sector arises near the base of the
radius, gradually diverging from it up to the junction of the middle and outer
thirds where a single branch is given off the latter taking a position parallel with
the main stem of the radial sector and entering the middle of the wing-apex
Eleven mm further out and within 7 mm of the apex a second branch is
given off, which lies evenly between the main stem and the first branch   The
median vein arises so close to the radius as to appear united with it   Like the
radial sector it does not divide until it reaches the junction of the middle and outer
thirds of the wing   Here the first inward branch arises and at equal distances
further out are given off two more, the first two each fork in the middle of their
length the third remaining undivided, the median ends therefore on the inner
wing-margin in six divisions   The cubitus consists of two, and possibly three,
main stems   Owing to the base of the wing being broken away these stems are
not seen to be in actual union   Little doubt can exist as to the union of the first
two, but if the third vein is a branch of the cubitus, it can only join the other two
by a strong outward curvature   This third vein may be the first anal, although its
manner of division is similar to that of the cubitus, in this respect agreeing with
what is seen in the first anal of *Lithosialis brongniarti*   The next vein is parallel
with the one succeeding, which is undoubtedly anal, while it shows an increasing
basal divergence from the second cubital   For this reason I have regarded the
third vein as the first anal   The first branch of the cubitus diverges widely from
the median, passing obliquely to the inner margin of the wing which is reached
just beyond the middle third   Owing to the great divergence of the first cubital
branch from the median, the area between the two, at the point where they first
branch, is very wide—almost twice the width of any other area

The first cubital vein gives off three outward twigs   The second is parallel
to the first as far as its division into two equal twigs   The next vein is that to
which I have already alluded as a possible third branch   It is widely separated

7

from the undoubted cubitus, and by reason of its two forward branches resembles the cubital elements in front of it.   Its wide separation from the second branch of the cubitus supports the view that it is the first anal vein.   The undoubted anal veins are two in number, undivided, and passing backwards very obliquely to the wing-margin.   More anal veins may have been present, but it is very doubtful.   The wing-surface is much wrinkled across both the length and breadth, and traversed by numerous cross-nervures, which are most evident in the median area.

Affinities.—The wing is typically Palæodictyopteroid, and in the development of the radius, radial sector, and cubitus, shows an affinity to *Lithosialis brongniarti*.   It differs markedly from that species, however, in the great length of the main stems before divisions arise.

Dr. Tillyard, to whom I have shown my enlarged drawings, is of opinion that *Pruvostia* is allied to *Pseudotœnigea* rather than to the Lithomantidæ.

Family BREYERIIDÆ, Handlirsch

1906   Handlirsch, Die Fossilen Insekten p. 95.

Wings markedly triangular, with broadened bases.   Costa, subcosta and radius brought closely together in the outer part of the wing; the median with few divisions.   Cubitus and anal veins directed almost straight inwards, at right angles to the length of the wing.

Genus **BREYERIA**, Borre

1875   *Breyeria*, Borre, Ann. Soc. Entom. Belg. vol. xviii p. 40.
1908   *Stobbsia*, Handlirsch, Die Fossilen Insekten p. 1348.

Generic Characters.—Wings two and a half times as long as wide.   Outer wing-margin straight, apex curved inwards.   Inner wing-margin strongly convex.   Costa marginal, subcosta parallel and joining the radius near the wing-apex.   Radius giving off the radial sector near the base of the wing, and reaching the apex undivided.   Radial sector subparallel to radius.   Median boldly curved inwards almost at a right angle.   Cubitus less curved than the median.   Anal veins few.   Interstitial neuration of two kinds, that of the intercostal and radial sector areas of short straight nervures, and that of the remaining areas of irregular thin nervures which tend to anastomose into a loose meshwork or reticulate arrangement.

Breyeria woodwardiana (Handlirsch)   Text-figures 12, 13

1903   Allied to *Lithomantis carbonarius*, Stobbs, Geol. Mag. (4) vol. x p. 521.
1906   Palæodictyopteron, sp., Handlirsch, Die Fossilen Insekten p. 126.

1906   *Lithomantis carbonarius (?),* Woodward, Geol Mag 15, vol iii p 26 fig 1
1908   *Stobbsia woodwardiana* Handlirsch Die Fossilen Insekten p 1348 text-fig

Type —Greater part of a left hind-wing Mr J T Stobbs Collection

Horizon and Locality —Peacock marls overlying the Peacock Coal, and near the
top of the workable Coal Measures, Foley, near Longton Staffordshire

Specific Characters —Radial sector dividing in its distal half into fine inwardly
directed twigs which end on the distal third of inner margin Median and cubitus
with few widely spaced branches occupying greater part of inner margin
Anal veins about three to four in number

Description —The specimen has not been available for examination, and my
observations are based on the figures published by Woodward and Handlirsch It
is probable that neither of these figures is wholly correct, Handlirsch pointing out
what appears to be an obvious error by Woodward in the character of the cubitus

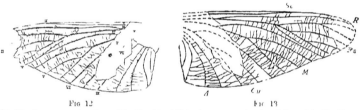

Fig 12 Fig 13

Fig 12 —*Breyeria woodwardiana* (Handlirsch) left hind wing as restored and figured by Dr H
Woodward under the name of *Lithomantis carbonarius ?* natural size —Coal Measures (Peacock
marls overlying the Peacock Coal) Foley near Longton, Staffordshire Mr J T Stobbs Collection
Fig 13 —*Breyeria woodwardiana* (Handlirsch) same left hind wing as restored and figured by Dr A
Handlirsch under the name of *Stobbsia woodwardiana* nearly natural size   *1* = IX and *Cu* =
VII, cubitus *M* = V median *R* = III radius *Rs* = IV radial sector *Sc* = II, subcosta *1,* costa

vein and the position and mode of division of the radial sector being doubtful
Notwithstanding this difficulty, I have by means of an old plaster cast been able
to satisfy myself upon the more essential details

The fragment is 50 mm long and 20 mm wide, and consists of the greater
part of a left wing, of which the base and a portion of the basal third are stated
to be obscured by a pinnule of *Neuropteris* The maximum width is across the
anal area, beyond which the wing rapidly narrows owing to the forward direction
of its inner margin The costal vein is marginal and strong, passing in an almost
straight line distally until joined by the subcosta, after which it curves gradually
backwards into the wing-apex The subcosta is parallel with the costal margin
over more than two-thirds of the wing and then joins the costa The base of
the radius is not shown in the figures, except a very small portion in front of the
point of origin of the radial sector the latter arising in the basal third of the wing,
beyond the radial sector the radius passes out to the wing-apex, keeping parallel
with the costal margin

The radial sector presents several difficulties. In neither of the two published figures is this vein depicted as we might expect. In all other respects the wing agrees remarkably closely with those of *Breyeria* and *Megaptilodes* where the radial sector sends off inwards a series of simple branches and runs out fairly parallel with the radius to the wing-apex. In place however, of the radius passing straight outwards, it is represented as dividing into two branches in the distal fourth of the wing, the outer branch forking once and the inner forking twice. The inner branch of the radial sector diverges widely from the outer and its three divisions go to the inner side of the wing-apex. A small branch is shown by Woodward as joining the radial sector to the first branch of the median while the same branch is shown by Handlirsch as coming off from the radius immediately in front of its division into two and passing down towards the inner margin between the inner branch of the radial sector and the first branch of the median but not uniting to the latter. The median vein arises near the radius and sweeps out in a bold curve to the middle of the inner margin giving off three outer undivided branches. Dr Woodward, in his restoration of the base of the wing, has inadvertently indicated the main stem of the cubitus is joining the median. This is corrected in Handlirsch's drawing. The cubitus consists of a strongly curved stem giving off two branches but only the inner marginal portions of the veins are present. The anal veins are three or four in number, and directed backwards at right angles to the length of the wing.

The interstitial neuration consists of feeble transverse nervures, which either pass irregularly across between the main veins or occasionally fork.

Affinities.—Dr Woodward doubtfully refers the wing to *Lithomantis carbonarius*, Woodw. Handlirsch, in the earlier part of his work 'Die Fossilen Insekten' (p. 126), classed it as a "Palæodictyopteron" only, and afterwards owing to its supposed relationship to *L. carbonarius*, and its evident likeness to *Lithosialis* and *Hadiromma* established the genus *Stobbsia* for it, placing the species in the Lithomantidæ. There are however certain features of the wing which militate against his view. The close apposition of the costa, subcosta and radius are in marked contrast to the condition in that family where these veins are widely spaced, and where there is also a very wide intercostal area. The radial sector is also more complex. In those details in which the wing departs from the Lithomantidæ, it approaches the characters of the genera *Breyeria* and *Megaptilodes*. The resemblances to *Megaptilodes brodiei*, Brong. and *Breyeria borinensis*, Borre, are very close, so far as can be determined by the distal fragment of the former wing and the more than three-fourths of the latter wing. The main difference between this wing and those of *Breyeria borinensis* and *B. harklani* is in the character of the subcosta which in the latter two species joins the radius. Whether the published figures of the Staffordshire wing are correctly drawn in this particular we do not know, and as we have already seen that these figures are wrong in other details, we are compelled to assign the specimen to the family Brevernidæ, and doubtfully to the genus *Breverna*, with which it seems in agreement

Family SPILAPTERIDÆ (Brongniart), Handlirsch

1906 Handlirsch Die Fossilen Insekten p 101

Radial sector more or less branched, median divided into two main branches, the outer much divided, cubitus with an outer branch sending numerous twigs to the inner margin Intercostal area occupied by a series of straight cross-nervures

The family Spilapteridæ founded by Brongniart in 1885, has been re-defined by Handlirsch and Lameere The latter extended the group to include species which Brongniart had placed partly in the family Platypteridæ and partly in Protephemeridæ All the forms thus brought together by Handlirsch agree in the possession of a typical palæodictyopteroid neuration and the general characters enumerated above

Lameere (Bull Mus Hist Nat Paris, 1917, no 1), who rejects Handlirsch's views and classifications (see above, p 15), has remodelled the family and given it a new significance He is of opinion that the genera *Lamproptilia*, *Epitelle*, *Becquerelia*, *Palæoptilus* *Componeura*, *Spilaptilus*, *Homaloneura* *Graphiptilus* and *Spilaptera* form a natural family the Spilapteridæ, in which a progressive evolution in the longitudinal venation can be observed The family is regarded as linked to that of Megasecopteridæ through the genus *Becquerelia*, and to the Protephemeridæ through the genus *Apopappus* The three families are then grouped in his new order Ephemeroptera Such a classification is based on the belief that a perfect evolutionary sequence can be made out Unfortunately, in presenting this classification, Lameere gives only a summary of his reasons and evidence and it is not possible to criticise his argument It is to say the least, very doubtful if, in the present state of knowledge, we can judge relationships always correctly, while the sequence of evolution is still more difficult

Handlirsch has acknowledged that he is unable to undertake any division of the family Spilapteridæ as he understands it, and is content to await the discovery of the bodies of these insects for a fuller knowledge of the family. As the more rational view, Handlirsch's definition of the family is adopted

Genus **SPILAPTERA**, Brongniart

1885 *Spilaptera*, Brongniart Bull Soc Amis Sci Nat Rouen [3] ann xxi, p 63

Generic Characters—Insects closely resembling *Palæoptilus*, fore-wings narrower basally than the hind-wings, body slender

Spilaptera sutcliffei, Bolton   Plate III, fig 3, Text-figure 11

1917 *Spilaptera sutcliffei*, Bolton Quart Journ Geol Soc vol lxxii p 53, pl iv fig 1, and text fig

Type —Basal third of a wing Manchester Museum (no L 8197)

Horizon and Locality —Middle Coal Measures (grey-blue shales 135—180 feet above the Roviey or Arley Mine), Sparth Bottoms, Rochdale, Lancashire

Specific Characters —Subcosta parallel with the costal margin; median vein dividing near the base into two branches, of which the outer forks just beyond the point of origin of the radial sector Cubitus a large and much-divided vein Anal veins few in number

Description —This specimen was formerly labelled " *Stenodictya lobata* " and is one of three recorded by Sutcliffe, Baldwin, and others in their papers on the fossils found at Sparth Bottoms The remaining two are described (p 37) under

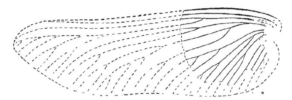

Fig 11—*Spilaptera sutcliffei* Bolton restoration of left wing natural size —Middle Coal Measures (shales above the Roviey or Arley Mine) Sparth Bottoms Rochdale Lancashire Manchester Museum (no L 8197)

Megnoptera tuberculata, Bolton The specimen consists of the basal third of a wing lying on the median plane of a small irregular micaceous sandy nodule The finer structure of the wing has not been preserved owing, no doubt, to the coarse grain of the matrix The chief veins of the wing are robust structures and these are fortunately well marked and clear The wing-fragment is 35 mm in length along the outer margin and 27 mm in greatest width but as the inner margin is broken away and lost the total width of the wing exceeded this and may have been over 30 mm It belongs to a left wing and when complete must have been at least 90 mm in total length The perfect insect must therefore have had a spread of wing of nearly 200 mm, or about 8 inches

The outer margin is feebly convex the subcosta fairly parallel, sunk in a groove and gradually approaching the outer margin as it passes towards the wing-apex The rate of approach is so gradual that the junction must have been far out near the wing-apex The intercostal area is crossed by a series of straight nervures which are oblique in their course outwards

The radius is strong, standing up above the surface, and almost parallel with the subcosta. The radial sector arises from the radius at about 22 mm from the base of the wing. The median vein, owing to its inward direction away from the radial sector, is better shown. It forks very low down into two equal branches, both of which are widely spaced. The outer branch divides almost opposite the point of origin of the radial sector while the inner branch forks much nearer the base of the wing, the innermost branch curving sharply back from its fellow, so that the area between them is wide almost from the commencement.

The general direction of the branches of the median is such that they would reach the outer half of the inner margin of the wing.

The cubitus consists of two main stems, the basal union being missing; the outer branch passes, after twice forking, in a double curve down to the inner margin of the wing, the inner branch being directed more directly backwards, and forking three times in the basal third. Fragments of four anal veins are distinguishable; the first dividing near the middle of its length into two equal branches. The interstitial neuration seems to have consisted of comparatively few straight and well-spaced nervures some of which can be seen crossing the wide area between the innermost twig of the median and its fellow. No others are visible, except those in the intercostal area already mentioned.

Affinities.—Notwithstanding the fact that only about one-third of the wing is known, it is yet possible to determine the generic characters with a reasonable degree of accuracy. The mode of division of the radius and median, the manifest importance of the latter, the wide area between the median and cubitus, and the few widely spaced cross-nervures, are all typical Spilapteroid characters. The wing cannot be confused with that of *Stenodictya*, in which the median is much less developed, and the interstitial neuration closely reticulated. The only other genus to which it might be referred, that of *Becquerelia* is distinguished by a union of the anal veins which in *Spilaptera* and in this specimen are distinct.

Spilaptera has hitherto comprised only three species. From *S. packardi*, Brong., the specimen differs by the subcosta being parallel with the outer margin, by the equal separation of the subcosta, radius and median in the basal third of the wing; and by the division of the median into two branches close to the base of the wing. *S. libelluloides*, Brong., of which the distal half of the wing only is known has a much feebler cubitus, which divides by forking further out than in *S. sutcliffei*. *S. tenuis* Brong. was established on a fragment much similar to that of *S. libelluloides* in which the subcosta is a short vein and the radial sector arises nearer the middle of the wing.

Family LAMPROPTILIDÆ (Brongniart) Handlirsch

1885   Brongniart, Bull Soc Amis Sci Nat Rouen 3e ann xxi p 67
1906   Handlirsch, Die Fossilen Insekten p 109

Fore- and hind-wing strongly marked   Veins of the anal and cubital groups numerous and directed obliquely backwards   Hind wings short and broad

Genus **BOLTONIELLA**, Handlirsch

1885   *Lamproptilia*, Brongniart Bull Soc Amis Sci Nat Rouen, 3e ann xxi, p 63
1919   *Boltoniella*, Handlirsch Revision der Palaozoischen Insekten p 21

Generic Characters —Fore-wings two and a half times as long as wide, hind wings as long as wide   Outer margin of former curved, of hind-wings almost straight   Apex of wings rounded   The fore-wings are almost elliptical, and the hind-wings rectangular in shape   The costa is marginal and the costal area somewhat narrow   The subcosta joins the costa far out   Radius simple, the radial sector arising near the base of the wing and giving off three forked branches which pass obliquely to the inner half of the wing-apex   Median dividing into two main branches, the inner twice as much divided as the outer   Cubitus with two main branches each breaking up into numerous twigs   Interstitial neuration consisting of widely spaced straight cross-nervures

Handlirsch, who has formed this genus, admits that it reminds one in many ways of *Lamproptilia*, but by reason of the small size and closer spaced cross-nervures he separates it from that genus

Boltoniella tenuitegminata (Bolton)   Plate III fig 4

1911   *Lamproptilia tenuitegminata* Bolton, Quart Journ Geol Soc vol lxvii, p 170, pl x, fig 6

Type —A right hind-wing, in a small fragile block of brown mudstone or shale, crowded with plant-remains, Museum of Practical Geology Jermyn Street London (no 24509)

Horizon and Locality —Coal Measures (No 2 Rhondda Seam, base of the Pennant Series)   1½ miles north-east of Resolven Station, Glamorganshire

Specific Characters —Broad wings of great tenuity and blattoid-like character Principal veins much divided, and fading out on the inner margin   Apex of wing bluntly rounded   Anal veins very numerous

Description —The type-specimen consists of a right hind-wing of considerable tenuity the underlying plant-remains being easily traceable through the texture of the wing   The greatest length is 20 mm, and the greatest breadth 7 mm   The neuration is so similar to that which I have observed in the hind-wings of blattoids from Commentry, France, that it is difficult to refrain from classing the wing as blattoid. The neuration is extensively branched and the wing-boundaries are not well marked, but the wing appears to have been somewhat quadrangular in outline with a sinuous inner margin, and the base much broader than is seen in the ordinary form of blattoid hind-wing. The outer margin seems to have been straight, a portion of it still remaining in the middle third. The two margins merge in a well-rounded apex. The subcostal area is narrow strap-shaped, and probably extended over the whole length of the outer margin. No traces of cross-nervures can be seen on it. The radius divides near the base, giving rise to a series of branches which curve inwards as they approach the apex of the wing. The course of the branches is irregular, the interspaces widening and narrowing, possibly owing to the wing having crumpled during deposition. The median divides into two branches low down, each of which is repeatedly forked, the final divisions becoming attenuated and untraceable before the inner margin of the wing is reached. The course of the cubitus is obscured by a reed-like plant, only two basal portions being distinguishable. The anal area is filled by a broad series of thread-like veins which sweep obliquely inwards in a fan-shape and occupy a large part of the inner margin.

The inner margin of the wing to a third of its total length is quite filmy, the veins crossing the area as faint shadowy lines. The distal two-thirds is more strongly impressed, while in the broad base of attachment the stems of the principal veins seem to have been more than usually robust. No trace of trans-verse nervures or reticulation can be seen.

Affinities. —The nearest analogues to this wing are, I believe, the forms described by Brongniart as *Lamproptilia grand'eury* and *L. stirum* ('Études sur le Terrain Houiller de Commentry, vol. iii [1893], pp. 467—70, pl. xxxv [19], figs. 7—9). It appears to be more closely related to *L. stirum* than to *L. grand'eury*, but is more quadrangular, and its costal area is broader. The anal portion of the wing is of greater tenuity, and occupies fully half of the inner margin.

Family BRODIIDÆ, Handlirsch

1906 Handlirsch. Die Fossilen Insekten, p. 113.
1919 Handlirsch. Revision der Palæozoischen Insekten, p. 73.

Wings in which the anal area is much reduced and specialised, the radius undivided, and the median, cubitus, and anal veins arched and directed back towards the inner margin.

From a re-examination of the originals in the British Museum, Handlirsch has concluded that the family does not belong to the Palæodictyoptera but to the Order Megasecoptera. This view seems at variance with his own definition of the

8

latter Order ('Die Fossilen Insekten,' p 312), where he states that the special distinctive characters of the Megasecopteridæ are the tendency to a reduction of the anal area of the wing, and the partial fusion of the median and cubitus with the base of the radius *Brodia* certainly agrees with the Megasecopteridæ in the reduction of cross-nervures, but the anal area is very long, and I have not seen a single case among the twelve or thirteen specimens in the British Museum where there is any clear sign of fusion between the median cubitus, and the base of the radius The bases of the three veins are closely brought together in the inner spatulate portion of the wing but this seems a natural result of the narrowing of the wing in this region

The family Brodidæ is more fully represented by individuals in the British Coal Measures than any other, thirteen specimens being known of *Brodia priscotincta* Scd alone To this family also seem to belong a series of small larval wings The nemation in these is immature, and the wings themselves are far too small and weak to have supported the body in flight The abdomen seems long and well segmented, the segments showing evidence of well-developed pleura, and they may also have borne tubercles The whole integumentary structure is thin and if at all chitinous, only feebly so, and but faintly outlined on the surface of the nodules The wing-shape is much like that of *Brodia* and the veins, so far as determinable, such as may reasonably be supposed to have developed into the typical *Brodia* type in the adult

While general evidence and association point to these wings being those of larval forms of *Brodia* and possibly of *B priscotincta* Scd , it is more judicious to retain them in a distinct group The same difficulties occur with other larval wings which cannot be allocated to known genera, and for which new generic names are not advisable I therefore class larval wings of unknown relationship as a separate group under the name of "*Pteronepiontes*," a name which has no classificatory value, but merely serves to indicate their larval nature When evidence is forthcoming of the generic and specific relationship of any member of the group it can be removed without the necessity of reducing a generic term which has passed into nomenclature to the rank of a synonym

I would allot the term "*Pteronepiontes*" to all larval insect-wings of all geological periods

Genus **BRODIA**, Scudder

1881   *Brodia*, Scudder, Geol Mag [2], vol viii p 293

Generic Characters —Wings spatulate in shape, three times as long as wide Outer margin feebly convex Costa and radius spinulose radius undivided radial sector and median vein with few divisions Cubitus a single vein   Anal veins few and widely spaced

Brodia priscotincta, Scudder   Plate III, figs 5, 6, Plate IV, figs 1–3, Text-figures 15, 16

1881   *Brodia priscotincta* Scudder, Geol Mag [2] vol viii, p 293 text-fig

1883   *Brodia priscotincta* Scudder, Mem Boston Soc Nat Hist, vol iii, p 215 pl xvii figs 3—7

1885   *Brodia priscotincta* Brongniart, Bull Soc Amis Sci Nat Rouen [3] ann xxi p 63

1893   *Brodia priscotincta*, Brongniart Faune Entom Terr Prim p 528, pl xl (24), fig 4

1906   *Brodia priscotincta* Handlirsch, Die Fossilen Insekten p 113, pl xii fig 13

1917   *Brodia priscotincta*, Bolton Proc Birmingham Nat Hist and Phil Soc vol xiv, pt 2 p 100, pl xii figs 3—1 text figs 2—3

1919   *Brodia priscotincta* Handlirsch Revision der Palaozoischen Insekten p 73 fig 83

1919   *Brodia Scudderi* Handlirsch, op cit p 74, fig 84 Brit Mus, no I 3879

1919   *Brodia petiolata* Handlirsch, op cit, p 74, fig 85 Brit Mus no I 2961

1919   *Brodia pictipennis* Handlirsch, op cit p 74, fig 86 Brit Mus, no I 2961

1919   *Brodia fasciata*, Handlirsch, op cit p 75, fig 87 Brit Mus no I 1557

1919   *Brodia nebulosa* Handlirsch, op cit p 75, fig 88, Brit Mus, no I 2961

Type —Incomplete wing, British Museum (Brodie Collection, no I 3896)

Horizon and Locality —Middle Coal Measures (clay ironstone nodule from the binds between the " Brooch ' and ' Thick " coals) Dudley, Staffs (Scudder gives the locality as Tipton)

Fig 15 —*Brodia priscotincta* Scudder diagram of a complete left wing showing position of tubercles and character of neuration natural size —Middle Coal Measures Dudley Staffordshire

Specific Characters —Outer margin of wing spinulose, almost straight, or feebly convex at the most   Subcosta remote from margin at base, and extending beyond the middle of the wing joining neither to the costa nor to the radius   Radius a strong vein, and spinulose   Radial sector arising in the first half of the wing, and giving off four branches   Median with two forward branches   Cubitus a single vein   Anal veins two   Inner margin strongly convex

Description —The Geological Department of the British Museum possesses no less than eleven examples of this species in addition to the type, while two more are in the Geological Collection of Birmingham University, one being in the Beale Collection, and the other presented by Dr Blake   The British Museum specimens are registered as follows I 1557, I 1567, I 3879, I 3896, In 18191, all from Coseley, I 2961, In 18433, In 18434, all from Tipton, In 18429 from Sedgeley, and In 18430 I 3866 I 2962 and In 18432   The following description and remarks are based on a study of all these specimens

The longest wing I have seen measures 58 mm in length, and we are not likely to be beyond the mark in estimating the total span of the insect as about 130 mm., or over 5 inches.

The outer or costal margin is straight, or feebly convex at most, over the greater part of its length, curving distally into the wing-apex. Close to the base it swells out into a slight hump-like elevation which is seen in all cases where the wing has been broken off close to the body. The whole of the outer margin bears a dense series of minute, conical, sharply pointed spinules of a black colour. These are arranged in two rows on the proximal half, with the points of the spinules directed towards the wing-apex. The bases of the spinules are expanded, and give a doubly corded or moniliated appearance to the wing-margin. Tillyard regards these spinules as modifications of large hairs which he has termed " macrotrichia," and I see no reason to dissent from his view.

The subcosta at its origin is widely spaced from the outer margin, and passes out beyond the middle of the wing, gradually approaching, but failing to reach it. The subcosta is a strong, straight vein, the greater part elevated above the level of the wing-membrane and distally flattening into it and disappearing.

The radius is a strong vein, more convex than the outer or costal margin and therefore more widely separated from it at either end than in the middle. The basal portion is parallel with the subcosta for the whole length of the latter, it then becomes parallel with the outer margin for a short distance, and afterwards curves into the wing-apex. A single row of spinules can be distinguished along the whole vein in some specimens, and gives a slight moniliation to the vein-surface. The radial sector is well marked, and comes off from the radius about the point of origin of the first forward branch of the median. The two veins are close together for some distance, but in the distal half of the wing they become parallel, the interspace being equal in width to that between the radius and the outer margin. The radial sector gives off four inward branches, which end on the inner side of the wing-apex. The first branch arises just beyond the middle of the wing, at an acute angle, and is separated from the second by an interval which is double the length of that separating the second and third. The fourth branch is very short, and so close to the margin as to be absent in some specimens.

The median vein for the first quarter of its course lies in the middle line of the wing, and then bends inwards in a wide curve to the distal third of the inner margin. It gives off two outward branches, both of which are larger and stronger than those of the radial sector. Both branches have the same sweeping curve possessed by the branches of the radial sector, and are parallel with the latter, while the main stem becomes almost straight.

The cubitus is a single vein, not united at the base to the median, and passing obliquely to the margin. Some wings have broken off so far out from the base that the cubitus appears to join the wing-margin at the junction of the middle and distal thirds. A striking feature of the cubitus is its isolated position upon the margin, for while the interval between it and the stem of the median is much wider than any other in the fore-part of the wing the interval between the cubitus and the next vein is nearly twice as wide There is, in fact, a progressive widening between the veins as they are traced from the apex to the base of the wing, the areas between the median and its branches being wider than those enclosed by the branches of the radial sector

Two somewhat dissimilar anal veins are present The first has a wide curve to a point far out on the margin, sometimes giving off a short branch, while the second vein is much shorter and joins the margin at an acute angle In some specimens this vein is seen to give off two, or even three short oblique branches

The shape of the wings is very similar to that of the wings of mosquitoes (*Anopheles*), and they bear evidence of having been folded in a plicate or fan-like fashion along their length The first two folds are united at the wing-base, and pass out along the radius and the first outward branch of the median The third fold lies along the line of the cubitus vein The degree of plication which a wing retained when silted up modifies considerably the apparent distance between the several veins and their branches, and at times hides important junctions In one example (Brit Mus, no In 18431) the wing was well flattened out before being buried, and the origin of the veins and their true position can now easily be determined This specimen shows that the radial sector arises much nearer the wing-base than the first outward branch of the median — a feature not usually shown in the remaining specimens

The general build of the wing is such that the bases of the costal and subcostal veins on the outer margin, and those of the anal veins on the inner, must have served as the main support to the distal two-thirds of the wing, the latter consisting mainly of the radial sector and its branches, the median, and the distal half of the cubitus Flight must have been mainly maintained by the action of this more distal expanded area, while the strain of movement would fall across the narrow neck-like base of the wing, and may ultimately have led to fracture and the loss of the wings It is quite possible, also, that these insects were capable of finding food among the decaying vegetation of the coal forests, and thus prolonging life for a considerable period after the wings were lost Such a presumptive sequence of events would account for the total absence of any trace of the bodies or legs, although the wings of this species are more numerous than any others in the Coal Measures and preserved in good condition

Colour Bands —Scudder mentions this species as the most striking instance among Palæozoic insects of the preservation of " colour bands," and as he states, some wings show three broad irregular belts of dull umber-brown colour across the wings Close examination of these ' colour bands ' in the type-specimen, and in other examples, leads us to doubt the correctness of his view In all cases where the ' colour bands " do not show on the wings the areas appear to be totally destitute of any traces of the wing-membrane, and the course of the veins only is shown across the matrix. It is much more likely that the 'colour-band' effect has been produced by conditions of preservation owing to the wing-membrane being destroyed in those areas which do not show colour.

Affinities.—Although Scudder founded both genus and species, he did not attempt any diagnostic description of either, confining his efforts mainly to a discussion of relationships. His figure is unusually poor and adds nothing to the text. Scudder's general conclusion was that the wing was nemopteroid in character, but "refusing to affiliate closely with the restricted families of the present day." A manuscript note in Brodie's handwriting placed with the type-specimen would seem to show that Scudder's views acquired greater definition later. Brodie writes under date, February, 1880 "I sent this wing to Mr Scudder, and he supposes it to belong to the white ants (Termitidæ), or close to the group of which Goldenberg's *Dulyoneura* is the best type." The first detailed descriptive note of the species was published in 1893 by Brongniart, with an

Fig 19.—*Brodia priscotincta*, Scudder, immature wing, twice natural size.—Middle Coal Measures (clay ironstone nodule from bands between the Frooch and Thick coals) Coseley, Staffordshire. Madeley Collection, Brit. Mus. (no I 2966) *A*, anal, *C*, costa, *Cu*, cubitus, *M* median *R*, radius, *Rs* radial sector, *S*, subcosta

excellent enlarged drawing. Brongniart very doubtfully assigned the wing to the Protodonata, and "alongside the Campyloptera."

Immature Wings.—The collection of insect-remains which I describe later (p 67) under the name of "*Pteronepionides*" were found at the same horizon and localities as *Brodia priscotincta* and it is fairly certain that some of them are immature forms of this species. It will be noted that these larval forms are broad-bodied and well segmented, and with lateral outgrowths of a pleura-like character upon the abdomen. They indicate that *Brodia priscotincta* went through a progressive metamorphosis, the rudimentary wings gradually developing as the insects lived as ground-feeders among dank and rotting vegetation.

One of these wings (Pl IV, fig 3) in the Madeley Collection in the British Museum (no I 2966) is an impression 18 mm in length by 4 mm in maximum breadth, contained in a small grey ironstone nodule. As is usually the case with these grey nodules, the details of the wing are much obscured by the matrix, and the precise method of division of the veins is far from being clear.

The outer margin is regularly convex and formed by the costa. It gradually merges into the expanded and well-rounded wing-apex. The subcosta extends the whole length of the wing, ending in the apex, and being parallel to the costal margin. A wide area separates it from the next vein

The course of the radius is similar to that of the subcosta. The point at which the radial sector arises cannot be determined, but it lies in the base of the wing

The radial sector is parallel with the radius over the greater part of its length, and gives off three inward branches in the distal third of the wing, all of which end on the inner side of the wing-apex

The course of the inner third of the median is indistinguishable. The main stem reaches the middle of the wing before it gives off the first outer branch, which bends round and becomes parallel with the inner branch of the radial sector. The second branch arises a little further out, and at a more acute angle than the first, passing down to the inner margin of the wing midway between the first branch and the main stem

The cubitus is a long undivided vein which passes, first in a curve and then in a straight line to the inner margin

Lying inward to the cubitus are traces of another undivided vein, which may represent the anal. It bends inwards more rapidly than the cubitus, and the interval between the two veins is very wide. The inner margin of the wing is sigmoidal in outline.

The general characters of this immature wing are unmistakeably those of the genus *Brodia*, and only the absence of spinules on the costal margin and radius, and the lack of a fourth branch to the radial sector, distinguish the specimen from adult wings of *B. priscotincta*. These details are not of specific value, and may be due to the immature condition of the wing which is but one-third the length of that of a normal *B. priscotincta*

The general characters of the wing are much like those of *Brodia priscotincta* (*proens*), and the specimen may be a slightly older larva of that form. The outline of the wing is much the same but the branching of the principal veins is better shown and their apical curvature less pronounced

The wing also lends considerable support to the belief that these insects did not possess a resting or pupal stage, but that the metamorphosis was regularly progressive

Brodia priscotincta, Scudder, *forma proens* Bolton   Plate IV, figs 1, 2, Text-figures 17, 18

1919   *Brodia nympha*, Handlirsch, Revision der Palaeozoischen Insekten p 76, fig 90

Type.—A pair of wings, one almost complete, the other showing the apical half only; British Museum (Johnson Collection, no 1 1563)

Horizon and Locality.—Middle Coal Measures (binds between the ' Brooch ' and ' Thick coals) , Coseley, Staffs

Description.—The fossil is contained in a brown ironstone nodule the whole wing having a length of 21 mm., and a maximum diameter of 4 mm   It is membranous, and oblanceolate in shape   The outer wing-margin, of which only the distal half is preserved, is flatly convex and curves inwards, meeting the inner margin in a rounded apical angle   The inner margin is feebly concave   The wing has three longitudinal folds, and the extreme tenuity of the integument is accompanied by a corresponding thinness of the veins   The greater portion of the costa and the whole of the subcosta are missing   The outer portion of the radius is present. It is parallel with the outer margin, and curves round into the wing-apex   The radial sector is parallel with the subcosta, and curving round into the apex is lost, by reason of its tenuity   Lying inward to the main stem of the radial sector in the distal half of the wing are two long veins, and traces of two others, all following the same course, and curving to the inner margin   The two long veins

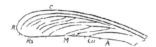

Fig. 17.—*Brodia priseolineata*, Scudder   *forma priseus* Bolton   diagram of wing nervation two-and-a-half times natural size.—Middle Coal Measures (binds between the "Brooch" and "Thick coals) Coseley, Staffordshire.   Johnson Collection Brit Mus (no I 1563)   Lettering is in Text-figure 16 p 62

appear to arise from the radial sector   in which case the two of which traces only are seen would do so also   The radial sector therefore seems to give origin to four inwardly directed parallel branches   The whole course of the next vein is not clearly determinable   It passes to just beyond the middle of the inner margin in an almost straight oblique course, and gives off a single forked outer branch parallel with the fourth branch of the radial sector   This vein can only be the median   The next vein is the cubitus   It is undivided, and goes to the middle of the inner margin   Anal veins are only indicated by feeble traces of a single undivided vein, which apparently reached the margin midway between the cubitus and the base of the wing

Another similar wing from the same horizon and locality in the British Museum (no I 1564) is 17 mm long and 5 mm wide, and lies on the surface of a split nodule of dark-brown ironstone   The wing-membrane is very thin and forms a slight glaze on the otherwise granulated surface—a feature which has made the details of structure difficult to determine   The costa is marginal the outer margin convex, and gradually curving into the rounded apex of the wing   The inner margin is slightly convex distally and straight proximally

The subcosta is a feeble vein whose course cannot be traced with certainty beyond the middle of the wing. It is close to the margin and parallel with it for its whole length. The radius is well marked, sunken, and also parallel with the margin except distally, where it is more inwardly curved. It ends on the apex in a short fork. The radial sector arises near the base and is parallel with the radius. It appears to give off three to four branches, the first arising very near the base of the wing. The whole course of this branch can be traced, but of the middle two only faint traces are left in the region of the wing-apex. The fourth branch is very short and corresponds in position with the last branch of the radial sector in *Brodia priscotincta*. The median vein forks low down into two equal branches, which reach the middle of the inner margin.

The cubitus is a long, undivided vein passing out almost to the middle of the inner margin, and separated from the next vein, which seems to be the first anal. The course of the latter is more oblique to the inner margin than that of the cubitus.

In its general character this wing is distinctly similar to that of *B. priscotincta*, although it lacks a second branch to the median, and the first anal does not seem to be forked.

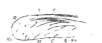

Fig. 18.—*Brodia priscotincta*, Scudder. *forma juvenis*, Bolton. diagram of wing neuration, twice natural size.—Middle Coal Measures (bands between the "Brooch" and "Tuck" coals). Coseley, Staffordshire. Madeley Collection, Brit. Mus. (no. I 15911). Lettering as in Text-fig. 16, p. 62.

Affinities.—Handlirsch regards these as nymphal wing-sheaths, but their extreme tenuity militates against this view. He is probably wrong in referring *Brodia* to the Megasecoptera, but right in regarding these wings as those of heterometabolous insects, and not holometabolous as Lameere supposed.

Immature though the wings undoubtedly are, the more nearly perfect example possesses an assemblage of characters which I believe points to its relationship. The shape of the wing, the character and course of the subcosta, the number and position of the branches of the radius, the simple forking of the median and the anal veins are all characters pertaining to the genus *Brodia*. The differences in detail between this wing and that of *B. priscotincta* are such as may be looked for between the nymph stage and the adult. Among previously described larval wings, the only type which seems comparable is *Lameereites curvipennis*, Handlirsch, based on four nymph-wings or wing-cases as Handlirsch has them, found in the Coal Measures (Pennsylvanian) at Mazon Creek, Illinois, U.S.A. (1911. Handlirsch, 'Amer. Journ. Sci.' [4], vol. xxxi, p. 375.)

In these wings the outer or costal margin is more strongly curved, and the costa and radius extend over the bluntly curved apex down to its junction with the inner margin. The succeeding veins arise nearer the base of the radius, only one

9

apparently being a branch of that vein, while the median and the cubital areas are occupied by numerous veins whose origins are not indicated

The differences between *Lameereites corripennis* and this specimen are considerable. The balance of evidence is greatly in favour of an affinity with *Brodia priscotincta*, and the specimen may represent a nymph or larval stage of that species.

It is undesirable to attach a specific name to immature wings agreeing so closely with a known species. I propose to regard it as *B. priscotincta*, forma *juvenis*

Brodia furcata, Handlirsch   Plate III, figs 7, 8. Text-figure 19

1919   *Brodia furcata* Handlirsch Revision der Palaozoischen Insekten p 75, fig 89

Type —A left wing and impression showing the under-surface, British Museum (no I 2962)

Horizon and Locality —Middle Coal Measures (above the ' Brooch " Coal), Dudley and Coseley, Staffs

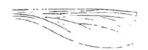

Fig 19—*Brodia furcata*, Handlirsch showing forking of the second branch of the median into two equal twigs natural size —Coal Measures (clay ironstone nodule from beds between the "Brooch" and "Thick" coals) Coseley, Staffordshire   Brit Mus (no I 2962)

Specific Characters —Radial sector reduced in area and possibly with three branches. Median having the second branch dividing into two equal twigs, both of which reach the margin. Median area enlarged

Description —This wing differs so much from the type-form as to be worthy of specific distinction. Its total length is 44 mm and the greatest width 12·5 mm. The base of the wing is much more nearly complete than usual and very narrow (5 mm) for a short distance, beyond which it widens by the development of the strong convex inner margin. The wing shows the usual plication, which fortunately is not continued into its base, so that the course of the veins in the latter is not obscured, as is so often the case in *B. priscotincta*. The outer or costal margin forms an almost straight line, and bears a double row of spinules, those at the base of the wing pointing inwards. No basal hump, as in *B. priscotincta*, is shown. The subcosta is widely spaced from the outer margin proximally. The radius and radial sector present no special features, and but one inward branch is present. Whether four branches arose from the main stem, as in the type-species cannot be determined owing to the loss of the apical portion of the wing. It is doubtful if such was the case, as the portion missing is not great. The outer three branches of the radial sector are usually 10—12 mm apart, so that if the wing possessed the same number of branches of the radial sector as in *B priscotincta*, its total length would have been 70 mm —an unusual length. The median vein is the most powerful of the whole series, and occupies a middle diagonal area equal in extent to the combined costal and radial areas. The first branch arises in line with the radial sector, and much nearer the base of the wing than in *B priscotincta*, and also lies so much nearer the cubitus that the area separating the two is but half the diameter of the area in the former species. This shortening of the main stem and its movement inwards has been brought about by the division of the second branch into two equal twigs, which pass out to the margin between the first branch and the main stem. The median vein therefore takes a larger share in the wing-structure than in *B priscotincta*, and the radial and cubital areas are correspondingly reduced. The cubitus presents no special features. Notwithstanding the shortening up of the area separating it from the median, the area between the cubitus and the anal veins is very large. The cubitus has suffered no displacement by the increased division of the median. The first anal vein is very long, passing well beyond the first third of the wing and dividing just before reaching the wing-margin. The second anal vein is two-thirds the length of the first, and bends inwards more gradually to the margin, giving off two short oblique branches in its basal half.

" PTERONEPIONITES "

Many larvæ of fossil blattoids have been recorded but very few of other groups. These larvæ may eventually reveal the changes undergone up to the adult stage, and the development of the neuration of the wings. In addition, the occurrence of larvæ in deposits is more likely to be indicative of habitat than the presence of adult wings as the inability of larvæ to fly and their lesser power of flotation would ensure inclusion in adjacent deposits. It is therefore necessary that their occurrence, and as much as possible of their structural appearance, be fully recorded. Any attempt to classify them under genera and species would rather retard than accelerate progress, and it seems advisable to record them under some term which will leave no doubt of their larval character. Handlirsch ('Amer Journ Sci' [4], vol xxxi, p 375, 1911) has already described larval " wing-cases " of a somewhat similar character under the generic name of " *Lameereites*," and placed them under the order Megasecoptera.

Had not Handlirsch given a generic value to the name " *Lameereites*,' it would have been possible to extend the use of his term to all larval wings. Failing this I would suggest the use of the word " *Pteronepionites* " for all larval wings which cannot be referred to a known genus, adding a specific designation when any larval wing presents features of a definitely recognisable character. Handlirsch restricts the name " *Lameereites* " to the wings of larval Megasecopteridæ, but recognises the close similarity between them and *Pteronepionites* ('Revision der Palaozoischen Insekten' p 76)

"Pteronepionites" johnsoni, sp nov   Plate IV fig 4   Text-figure 20

Type —Immature wing, 12 mm long and 4 mm wide, British Museum (Madeley Collection, no I 2967)

Horizon and Locality —Middle Coal Measures (bands between the "Brooch" and "Thick" coals), Coseley, Staffs

Description —The wing, like all these immature structures is of extreme tenuity and the finer details are masked by the coarse granular nature of the matrix composing the nodule   The outer margin is straight, curving backwards until it meets the inner margin in a bluntly-pointed apex   The inner margin is strongly convex, a distal infolding indicating that the wing was apparently never fully extended   Very feeble traces occur of a short straight subcosta, which

Fig 20— *Pteronepionites johnsoni* sp nov   diagram of wing neuation two and a half times natural size —Middle Coal Measures (bands between the "Brooch" and "Thick" coals), Coseley, Staffordshire   Madeley Collection, Brit Mus (no I 2967)

reached the margin near the apex of the wing   The radius is stout and straight and reaches the wing-apex, giving off a well-marked radial sector, which, diverging from the radius passes in a wide curve into the wing-apex   The median remains undivided for nearly a third of its length, and then bifurcates into two equal branches which assume a parallel position, and reach the outer third of the inner margin   The cubitus is represented by two veins, the first having a sigmoidal sweep and the second a simple curve   Both reach the middle third of the inner margin   No trace of anal veins can be observed

"Pteronepionites" ambigua, sp nov   Plate IV, fig 5

Type —A pair of larval wings still attached to the crushed and almost obliterated body, British Museum (Madeley Collection no I 2968)

Horizon and Locality —Middle Coal Measures (bands between the "Brooch" and "Thick" coals), Coseley, Staffs

Description —The insect lies in a fragment of a light grey ironstone nodule, and is not well preserved   One wing is whole but more than half of the second is missing and its apex is obscured by the matrix   The whole wing has a length of 6 5 mm and a width of 2 mm   Each is much thickened over the basal half and strongly ridged or furrowed along the line of a powerful vein or tracheal trunk (the wing is so immature as to render the latter possible), occupying the position of the radius and median veins. Traces of a third vein occupying the position of the cubitus are present on the inner half of the wing. The body-segments are numerous, 3—4 mm. in depth, and seem to have had pleura-like expansions. The region in front of the thorax is bent backwards and below the abdominal segments. The thoracic segments are larger and more robust than those of the abdomen.

The thinness of the body, and the difficulty of determining boundaries with satisfactory accuracy render all attempts at a more precise determination impossible. We can, however say with confidence that the remains are those of an insect possessing a long segmented body, an elongated head-region and wings carried upright over the back.

"Pteronepionites" lepus, sp. nov.   Plate IV, fig. 6

Type.—Remains of a larval insect, having a segmented body and two immature wings, in a flattened nodule of light-grey ironstone. British Museum (Madeley Collection, no. I 2969).

Horizon and Locality.—Middle Coal Measures (binds between the 'Brooch' and 'Thick' coals). Coseley, Staffs.

Description.—The impression of two wings is clearly discernible; the segments of the abdomen are less so. With oblique lighting six segments can be made out behind the thorax. The wings are slightly unequal in size, and the anterior is longer and thicker than the posterior wing; it is also somewhat infolded at the base. The differences between the two wings and their relative positions, would seem to indicate that they are the fore- and hind-wings of the one side.

The fore-wing has a length of 9 mm. and a width of 2.5 mm. while the hind-wing is 8 mm. long and 2 mm. wide. The fore-wing has a strap-shaped appearance, and the line of attachment to the body is its broadest part. A stout ridge, swollen at its junction with the body, traverses the greater part of the length of the wing, dying out before the apex is reached. The wing is too small for definite determination of this swelling, but it appears due to incomplete extension of the wing-membrane rather than to the presence of a vein. The distal fourth is flat and thin, and the apex well rounded. The outer and inner margins are parallel and undulated. A few faint and irregular lines may be indications of veins.

The hind-wing has undergone greater expansion than the fore-wing, and lies almost flat on the nodule. Both outer and inner margins are gently convex, the convexity of the inner margin being the greater. Feeble traces of a subcosta and of a stout vein which forks before the middle of the wing is reached, are distinguishable   The latter vein occupies the position I should assign to the radius No further traces of veins are visible

The two wings are separated by an interval of 3 mm at their bases   No definite details can be made out in the thoracic region   Behind the hind-wing are faint impressions of a series of abdominal segments. These are about three times as deep as wide and appear to have borne lateral spiny processes   There are also traces of tubercles

The head-region is only indicated by faint discolorations   As in the case of other examples of " *Pteronepionites* " we have met with, the remains are so filmy in character, and merge so much into the matrix, that it is impossible to define the outer limits of the various segments with absolute clearness, and no attempt can be made at a systematic determination of characters

The specimen is a larval form with wings not yet fitted for flight, but with a degree of differentiation in the fore-wings which indicates that they were thicker and less flexible than the hind-wings

The abdomen is long, wide, and well segmented the lateral expansions being not unlike those of *Euphoberia ferox*   So closely does the abdomen resemble the segmented body of a Diplopod, that in the absence of the wings we believe it would be readily classed as belonging to that group, and as possessing nothing in common with insects

It is difficult to resist the belief that these larval insects were capable of crawling about in decaying vegetation, and that their larval life was thus spent until by successive ecdysis the wings had acquired sufficient strength to lift the body from the ground, and enable the insect to fly

Family ÆNIGMATODINÆ Handlirsch

1906. Handlirsch Proc U S National Mus vol xxix p 683
1906 Handlirsch, Die Fossilen Insekten p 116

Wing strongly arched and broadly rounded at the apex   Anal area small, and not marked off from the rest of the wing   Subcosta reaching almost to the wing-apex radius simple, and radial sector with three divisions   Median with four branches   Costa represented by an oblique vein with a terminal fork, followed by three simple strongly curved anal veins   Interstitial neuration partly of regular cross-nervures, and partly of a polygonal network

Handlirsch founded this family on an incomplete wing in which the greater part of the outer border, subcosta and radius is missing   From the upper Middle Coal Measures of Mazon Creek Ill, U S A

Genus **ÆNIGMATODES**, Handlirsch

1906 *Ænigmatodes*, Handlirsch, Die Fossilen Insekten, p. 116

Generic characters as above

Ænigmatodes (?) regularis, sp. nov.   Plate IV. fig. 7. Text-figure 21

Type.—Fragmentary wing. British Museum (no. In. 18,604)

Horizon and Locality.—Middle Coal Measures (over the Barnsley Thick Coal). Monckton Main Colliery. Barnsley, Yorkshire.

Specific Characters.—Radial sector, median, and cubital areas occupying most of the wing, and all with well-spaced branches united by straight nervures, except in the marginal area between the median and cubital veins where they form a slight meshwork. Inner margin well rounded.

Description.—Little more than half of this wing is preserved, and lies on a fragment of hard grey bind associated with broken-up plant-remains. The greatest length of the fragment is 42 mm., and the width 15 mm. The length

Fig. 21.—*Ænigmatodes (?) regularis* sp. nov.—diagram of neuration of wing fragment, one-and-a-half times natural size.—Coal Measures (over the Barnsley 'Thick' coal), Monckton Main Colliery. Barnsley, Yorkshire. Brit. Mus. (no. In. 18,604). Lettering as in Text-figure 16, p. 62

of the complete wing was probably 50—60 mm., and the width 18—20 mm. The outer margin is missing, and of the subcosta and radius only portions of the basal third are present. These are separated by a narrow area crossed by short transverse nervures.

In the distal third of the wing is the innermost branch of the radial sector, which ends on the margin in a small fork. The line of fracture of the wing has closely followed the line of the radial sector, and a portion of another branch of the radial sector may have been present along the line of the extreme broken edge of the wing. The median vein sweeps in a convex curve to far out on the inner margin, giving off two well-spaced branches. The next two veins, one of which forks, are more oblique in their course, and may also belong to the median, but the basal curve rather indicates that these veins are cubital, as is the succeeding vein, of which only a small portion is left. A trace of an anal vein is present near the wing-margin. The interstitial neuration consists of well-marked transverse nervures, regularly spaced in the outer parts of the wing, and uniting in a loose meshwork in the wide area between the inner branch of the median and the cubitus

Affinities—The nearest approach to a wing of this character is that of *Enigmatodes danielsi*, Handlirsch, in which the greater part of the outer margin is also missing. The Yorkshire specimen which may be a hind-wing, is three times as long as *E. danielsi* and the veins have a stronger inward curvature. In the absence of more definite details, it seems best provisionally to refer the specimen to the genus *Enigmatodes*.

Genus **PSEUDOFOUQUEA**, Handlirsch

1906 *Pseudofouquea* Handlirsch, Die Fossilen Insekten p 125

Wings three times as long as wide. Cubitus with inner and outer branches. Anal veins, so far as known not united. Interstitial nervation of feeble cross-nervures except between cubitus and first anal where it is irregularly reticulate

Pseudofouquea cambrensis (Allen)   Plate IV. fig 8   Text-figure 22

1901 *Fouquea cambrensis* Allen Geol Mag 4, vol viii p 65 text-fig on p 66
1906 *Pseudofouquea cambrensis* Handlirsch, Die Fossilen Insekten p 125 pl xm fig 5
1916 *Pseudofouquea cambrensis* Bolton Quart Journ Geol Soc vol lxxii p 59, pl iv, figs 4—5, and text-fig

Type—A broken left fore-wing of which the two parts are preserved on fragments of black shale, one fragment bearing the basal part of the wing in the Museum of Practical Geology, Jermyn Street (no 7272) the other, containing the impression of the distal 28 mm of the wing in the Welsh National Museum, Cardiff (no 13120)

Horizon and Locality—Lower Coal Measures (top of the Four Foot Seam), Llanbradach Colliery near Cardiff

Specific Characters—Wings stout and obtusely pointed. Costa marginal and flatly convex. Subcosta reaching margin in the outer third. Radius parallel with subcosta and ending in apex of wing. Radial sector diverging widely from radius, and ending on the wing-apex in five divisions. Comstock regards it as a typical dichotomous radial sector, with an accessory vein on the second branch ('Wings of Insects' 1918. Radius and radial sector occupying all the wing-apex. Median dichotomously branched, with an accessory vein in the fourth branch. Cubitus widely divergent from median and giving off alternate twigs from its outer and inner sides those of the inner side being much the feeblest. Anal veins four in number. Interstitial nervation of feeble cross nervures, except between the base of the first anal vein and the cubitus where it is irregularly reticulate

Description—The total length of the wing is now 32 mm 9 mm having been

lost from its tip since it was measured by Allen who gives the total length as
41 mm   The greatest width is 15 mm

The subcosta is widely separated from the outer or costal margin in the base of
the wing, and gradually approaches and unites with it beyond the middle   The
radius is parallel with the subcosta throughout its length, and gives origin to the
radial sector basally   The radial sector diverges from the radius over the whole
of its course   It now shows but one inwardly directed branch, which forks at the
broken edge of the wing   Allen's figure indicates that two more branches were
given off, both being undivided   The radius and radial sector occupied almost the
whole of the apex of the wing

The median vein forks in the basal fourth, the outer branch again forking
before the middle of the wing is reached   The inner branch of the median diverges
almost in a straight line from the outer branch, and also forks and bears an
accessory twig upon the fourth branch   A feeble accessory twig appears to be
given off near the middle of the wing but dies out in the integument   The cubitus

FIG 22—*Pseudofouquea cambrensis* (Allen)   restoration of whole wing natural size   Lower Coal
Measures (top of the Four Foot seam)   Llanbradach Colliery near Cardiff   Basal portion of
wing in Mus Pract Geol (no 72721   Impression of apical portion of wing in the Welsh National
Museum (no 13 120)

is a remarkable vein, unlike that of any other fossil wing (Comstock, 'Wings of
Insects, 1918, p 106)   For nearly half its length it passes in a broad curve to
the inner margin giving off a series of alternate twigs upon its outer and inner
sides, those of the inner being weaker than those of the outer side   The two
outer twigs are strongly developed, while those on the inner side of the cubitus,
four in number, are weaker and shorter   The feeble continuation of the main stem
reaches the margin between the two sets of branches   Four anal veins are
distinguishable   The inner two arise from a common base, the outer two not
uniting   This is unlike the condition in *Fouquea* where the anal veins branch off
regularly from a single stem

The area lying between the first anal vein and the main stem of the cubitus is
very wide—much wider, indeed, than any other area

The interstitial neuration consists of weak cross nervures, except between the
base of the first anal vein and the main stem of the cubitus where it is irregularly
reticulate

Affinities — The characters of the cubital and anal veins definitely remove the
species from the genus *Fouquea*, and the cubitus, with its strong, anteriorly
directed twigs and its feebler inner series, is wholly unlike that of any other
insect, and would alone suffice to justify the generic rank given by Handlirsch.

10

So far I am in agreement with Handlirsch, but I regard the enlarged areas between the inner divisions of the radial sector and the cubitus, and between the cubitus and the anal veins as more suggestive of the Protorthoptera, notably *Thoronysis nagbelensis*, Ammon. More than this cannot be said, and *Pseudotonquea cambrensis* must be regarded provisionally as Palæodictyopterid, with a possibility of Protorthopterid or even Orthopterid affinities.

INCERTÆ SEDIS

Genus **ARCHÆOPTILUS**, Scudder

1881   *Archæoptilus* Scudder Geol. Mag. [2], vol. viii, p. 295

A wing of unusually robust type. Only one is known, consisting of not more than the basal fifth of a whole wing whose total length may have been 25.4 cm. to 35.5 cm. The fragment is too small for a correct determination of its systematic position, and has been referred to widely separated families by various workers.

Archæoptilus ingens, Scudder   Plate IV, fig. 9, Text-figure 23

1881   *Archæoptilus ingens* Scudder, Geol. Mag. [2], vol. viii, pp. 295, 300
1883   *Archæoptilus ingens*, Scudder, Mem. Bost. Soc. Nat. Hist., vol. iii, pp. 217, 223, pl. xvii, figs. 10—12
1885   *Archæoptilus ingens*, Brongniart, Bull. Soc. Amis Sci. Nat. Rouen [3], ann. xxi, p. 69
1885   *Archæoptilus ingens*, Scudder, Zittel's Handbuch der Palæontologie, vol. ii, p. 757
1893   *Archæoptilus ingens*, Brongniart, Faune Entom. Temps Prim., p. 498, pl. xxxvii, fig. 6
1906   *Archæoptilus ingens*, Handlirsch, Die Fossilen Insekten, p. 117, pl. xii, fig. 18

Type.—Basal fifth of wing, in counterpart, British Museum (no. I 9997)

Horizon and locality.—Middle Upper Coal Measures. Between Shelton and Clay Lane near Chesterfield, Derbyshire.

Specific Characters.—Wings very large, costa, subcosta, and radius broad and robust. The costal border spiny. Interstitial neuration of stout transverse nervures.

Description.—Only the basal part of the wing and its counterpart are preserved, having a total length of 43 mm., and a greatest breadth of 33 mm. Scudder's estimate of the length of the whole wing as 35.5 cm. is probably excessive.

Scudder (*loc. cit.*, 1881) thus describes the specimen. "All the principal veins are a millimetre or more thick, and the cross-veins of the upper interspaces are tolerably distant, stout, prominent, and generally simple. The marginal (costa) vein, forming the front (outer) border of the wing is studded with short oblique spines (= macrotrichia). The other veins lie at very different levels on the stone and below the interspaces mentioned, seem rather closely crowded, and much more curved sweeping downward, while the upper veins show little tendency to turn from a longitudinal course."

The great apparent width of the costa is caused by the formation of an expanded chitinous bar along its outer margin, the free edge bearing the spines described by Scudder. The costa, with its chitinous bar, the subcosta and the radius are so broad as to appear strap-like, are widely separated and the intervening areas are crossed by equally strong nervures. The costa can be distinguished on the cast from the frontal bar and appears as a narrow rounded vein

The subcosta is a very broad vein, crossed by oblique striæ which are directed outward from the upper edge of the vein. A broad interval (9 mm.) separates it from the costa, the area being crossed by stout, slightly oblique transverse nervures. The general direction of the subcosta is such that it must have reached the margin of the wing near the apex. At the base it is much enlarged, the

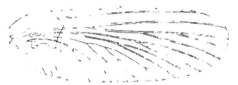

Fig. 23.— *Archæoptilus ingens* Scudder; drawn in of whole wing from Scudder's original restoration one quarter natural size.—Upper Coal Measures, near Chesterfield, Derbyshire. Brit. Mus. (I 8997)

enlargement probably indicating the attachment to the body of the insect. The expanded inner portion of the base is fused with the equally expanded base of the radius

The radius diverges from the subcosta in its outward course, and is even more enlarged than the subcosta. It is connected with the latter by a series of thirteen transverse nervures in the wing-fragment. Most of these are slightly convex, and two are united by lateral branching

The median vein was apparently closely apposed to, or united with the base of the radius, and is much less robust than the latter, the intervening area being narrow and crossed by short, thin, straight transverse nervures which do not appear to continue into the base of the wing. The cubitus is not so readily distinguishable, and diverges sharply inwards towards the inner margin. The anal veins are 4—5 in number, and are strongly curved inwards occupying not more than one-fifth of the inner margin of the wing. The first three may have been united at their base. The wing is marked by a series of folds along the lines of the principal veins

Affinities.—This remarkable wing-fragment has caused considerable conjecture as to its true character and relationship. Scudder in his second note (*loc. cit.*,

1883) somewhat vaguely placed it, 'with a strong degree of probability, in the same general group as some other Palæozoic wings." This reference can only be to the Palæodictyoptera. Later he published a restoration of the wing, and classed it with the Protophasmidæ. Brongniart at one time regarded it as a Dictyoneuron, and later as a "Nemorthopteron" of the group Sthenaropterida. Still later he placed it as a Neuropteron in the group Platypterida. Handlirsch considers that the costa is marginal, and that this character, with the sharp inward curve of the anal veins, justifies its inclusion in the Palæodictyoptera, although he does not attempt to indicate its allies.

A second species was described by Brongniart as A. lacasi ('Bull. Soc. Amis Sci. Nat. Rouen' [3], ann. xxi, p. 60, 1885), but this throws no light on the genus, nor does it appear to be generically referable to Archæoptilus.

The only details preserved which can be used in the determination of relationship are the spiny outer margin, the great width of the principal veins, the well-developed cross-nervures, and the strongly curved and numerous anal veins. Even these are too fragmentary for safe conclusions to be drawn in the absence of other material.

The general structure of the wing-fragment is to me more suggestive of the Protodonata than of the Palæodictyoptera, but the presence of well-marked anal veins discounts this view, unless we are prepared to accept the specimen as an early and archaic prototype of the Protodonata.

Order MIXOTERMITOIDEA, Handlirsch

1906   Handlirsch, Proc. U.S. National Museum, vol. xxix, p. 695, and Die Fossilen Insekten, p. 126.
1919   Handlirsch, Revision der Palæozoischen Insekten, p. 26.

Subcosta much shortened; radial sector arising close to the base, with 2—3 branches, of which only one forks. Median long, four-branched and suggestive of the Palæodictyoptera. Cubitus with 2—3 inward branches. Anal vein 3, simple. Cross-nervures strong, wide-spread and regular.

This is a provisional order established by Handlirsch to include two forms only—Mixotermes lugauensis, Sterzel, from the Coal Measures of Saxony, and Geroneura wilsoni, Matthew, from the Carboniferous of St. John, New Brunswick, North America. Both wings show clearly their Palæodictyopterous ancestry, but Handlirsch is uncertain whether they should be brought near to the Protorthoptera or to the Perlidæ.

Genus **GERONEURA**, Matthew

1889   Geroneura, Matthew, Trans. Roy. Soc. Canada, vol. vi, sect. iv, p. 57.

Generic Characters.—Wing three times as long as wide; apex obtusely rounded, subcosta short and joining outer margin before middle of wing. Radius and radial sector occupying almost the whole of the wing-apex. Median, cubitus, and anal divisions few. Interstitial neuration of stout cross-nervures at wide intervals.

Geroneura (?) ovata, sp. nov. Plate V, fig. 1

Type.—Portion of left wing, British Museum (Madeley Collection, no. I. 2965).

Horizon and Locality.—Middle Coal Measures (binds between the "Brooch" and "Thick" coals), Coseley near Dudley, Staffs.

Specific Characters.—Radial sector rising in outer third of wing, with three divisions. Median vein large, the two outer branches forking in line with origin of the radial sector, the third branch undivided. Cubital veins few, undivided, and reaching the distal part of the inner margin of the wing.

Description.—The specimen consists of the impression of the upper surface of the distal portion of a left wing, having a length of 32 mm., and a breadth of 20 mm. The impression lies on the surface of a thin flattened half-nodule of fine sandy grit, and is but faintly indicated. The total length of the wing was probably from twice to three times the length of the portion preserved, and its breadth may have been a little more than 26 mm. The outer margin is gently convex and curves into the broadly rounded apex. Very little is left of the inner margin, which also seems to have been convex. The distal portion of the costal margin is present for a length of 20 mm. There is no trace of the subcosta, so that this vein did not extend much, if at all, beyond the middle of the wing. The radius gives off the radial sector about the distal third of the wing, the two veins remaining almost parallel with the wing-apex. The radial sector gives off a single inward forked vein. The next three veins seem to be divisions of the median. The first two each give off an outer branch in line with the division of the radius and radial sector, and the outer of the two also forks before reaching the edge of the wing. The third vein is single for its whole length, but evidently united with the second a short distance outside the line of fracture of the nodule. The remaining four veins appear to belong to the cubitus. No anal veins are distinguishable. All the veins, with the exception of the small forks of the radial sector and first median, are parallel and widely spaced. They are united by a series of strong, straight cross-nervures placed widely apart. Notwithstanding the strength of the veins and of the cross-nervures, the smooth impression of the wing-fragment seems to indicate that the veins were not sunk below the general surface of the wing, as is usually the case.

Affinities.—The determination of the relationship of so small a wing-fragment would be difficult were it not for the unusual direction of the main veins, their mode of branching and the character of the cross-nervures. These characters are a special feature of the order Mixotermitoidea Handl.

The wing-fragment must be referred to this provisional order in the absence of knowledge of the whole wing-structure. The much-divided median vein is more comparable with that of *Geroneura wilsoni* Matthew, than with that of *Miotermes luganensis*, Sterzel, and is also correlated with a shorter subcostal vein, although in *G. wilsoni* that vein extends beyond the point at which the radial sector arises from the radius. An open series of cross-nervures is present in both genera as in this specimen, and both have the same well-rounded apex. The wing-fragment is suggestive of *Homoeistia occidentalis*, Dana, but has a less branched radial sector. I provisionally refer it to *Geroneura* with the specific name of *ovata*.

Order PROTORTHOPTERA, Handlirsch

1906   Handlirsch. Proc. U. S. National Museum, vol. xxix, p. 695 and Die Fossilen Insekten p. 123
1919   Handlirsch. Revision der Palaeozoischen Insekten, p. 28

Head large, with strong mouth-parts, and bearing long slender antennae; prothorax large and elongated, and the body strongly built. Legs either uniform in character and fitted for running or the hind-legs modified for leaping. Wings more specialised than those of the Palæodictyoptera and capable of folding on the abdomen when at rest, with the enlarged anal areas of the hind-wings doubled under owing to the formation of a fold between the anal area and the rest of the wing. The principal veins and their subdivisions not so strongly curved inwards as in the Palæodictyoptera.

Handlirsch established this order to include a series of insects intermediate in character between true Orthoptera and Palæodictyoptera, to which Scudder had previously given the name of Palæodictyoptera Neuropteroidea.

Genus **ÆDŒOPHASMA**, Scudder

1885   *Ædoeophasma*, Scudder, Geol. Mag. 3, vol. ii, p. 265

Generic Characters.—Large wings two-and-a-half times as long as wide; inner margin more convex than outer margin, and curving distally into the latter. Principal veins broad and flat in the basal third and diminishing in size distally. Subcosta and radius reaching the wing-apex. Median vein with two main branches, the outer with most subdivisions. Cubitus with two main branches each much subdivided. Anal veins numerous. Interstitial neuration of irregular nervures and a loose meshwork in the wider areas.

Ædœophasma anglica, Scudder   Plate V, fig. 2. Text-figure 21

1885   *Ædoeophasma anglica*, Scudder, Geol. Mag. 3, vol. ii, p. 266 and in Zittel's Handbuch der Palæontologie, vol. ii, p. 758, fig. 941

1906   *Ædœophasma anglica*, Handlirsch, Die Fossilen Insekten, p. 125, pl. xiii, fig. 4.
1916   *Ædœophasma anglica*, Bolton, Quart. Journ. Geol. Soc., vol. lxxii, p. 43, pls. iii, iv, and
   text-figure.

Type.—Greater part of a left wing in an ironstone nodule; Liverpool Museum (presented by Major Chambers in 1858).

Horizon and Locality.—Middle Coal Measures; South Lancashire (locality unknown, but the nodule so similar to those derived from the Ravenhead Railway Cutting that it may be from that section).

Specific Characters.—As generic characters.

Description.—The specimen was partly described and named by Scudder in 1885, and re-examined and figured by the present writer in 1916.

The wing lies in counterpart in a fine-grained ironstone nodule, and its total length as now exposed is 87 mm., its greatest breadth (across the middle) 40 mm. When whole, the wing was probably 100 mm. long.

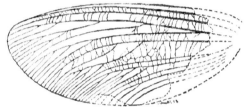

Fig. 21.—*Ædœophasma anglica*, Scudder: restoration of whole wing, showing the general character of the venation, natural size.—Middle Coal Measures; South Lancashire. Liverpool Museum.

The outer or costal margin is gently convex. The subcosta is a broad flat vein, gradually diminishing in width towards the wing-apex, which it just fails to reach.

The radius is an even broader vein than the subcosta, is also flattened in its basal third, and reaches the outer angle of the wing-tip, keeping parallel with the subcosta.

The median vein divides low down into two equal branches, the outer giving off four inwardly directed twigs. The first of these remains undivided; the second forks twice, and the outer and inner divisions of the second bifurcation again divide, so that the vein ends on the apical margin in six divisions. The remaining two branches are undivided. The divisions of the outer branch of the median occupy the greater part of the wing-apex.

The inner branch of the median does not divide until it has reached the apical fourth of the wing, where it gives off four twigs which pass inwards to the junction of the inner margin with the apex. Only the first of these twigs forks.

The cubitus has the same broad flattened basal portion which is so characteristic of the veins we have already dealt with. The main stem lies somewhat near the first branch of the median and remains parallel with it over almost the whole of its length. Inwardly it is separated somewhat widely from a slighter vein which, I believe, joined it near the base, and formed the first inward branch. The main stem sends off at the middle of the wing a strongly curved branch which bends first inwards and then outwards towards the apex, breaking up into five twigs before reaching the inner margin. The second of these twigs forks. A second undivided branch comes off a little further out, and a third very small one almost on the margin. The next two veins were probably united a little way out from the base, and their direction is such that the single stem from which they arose may have arisen, as suggested above, as the first inward branch of the cubitus. The outer of the two veins is undivided and reaches the inner margin beyond the middle of the wing. The innermost vein runs fairly parallel with the first along its whole length, giving off as it does so, four inwardly directed twigs, of which the first and fourth fork. The whole vein ends on the margin in six twigs. Four anal veins are shown, one only forking.

The interstitial neuration of the radial and median areas consists of straight or slightly curved nervures, placed at nearly equal distances. The very wide cubital and cubito-anal areas are filled by a loose meshwork, and a few irregular wavy nervures. The anal area is crossed by simple straight nervures.

Affinities.—Scudder was originally of opinion that this wing was related to *Meganeura* (*Dictyoneura*) *monyi*, Brong., representing a member of the group Protophasmidæ. Handlirsch removed the genus to the group of Palæodictyoptera incertæ sedis.

Scudder was undoubtedly mistaken in referring the wing to the Protophasmidæ, as a glance at the figure of *Protophasma dumasii*, Brong., will at once show (Die Fossilen Insekten, pl. xvi, figs. 1—2). Handlirsch did not see the specimen, and had to base his determination on a sketch of the wing which he considered "confusedly drawn." The latter probably accounts for the interpretation which he placed on the various principal veins. More recently I have been able to expose more of the structure, and diagnosed the wing accordingly. If Handlirsch's view were correct the radial sector would be of enormous proportions and occupy all the wing-apex. The base of the radius, so far as shown, is widely divergent from the base of the median—more so, in fact, than at any other part of the whole course of the radius and supposed radial sector. These veins therefore have come into union only at the actual point of origin of the wing. This may have been the case, but in my opinion the radius is wholly simple and undivided and no radial sector is present. The median and cubitus are large, much divided and take up the greater part of the wing-area, while the anal veins are few.

If this view be correct the wing is a very primitive example of the Protorthoptera, still retaining evidence in the costa, subcosta and radius of its Palæodictyopterod origin.

Bolton. Insects of Coal Measures.

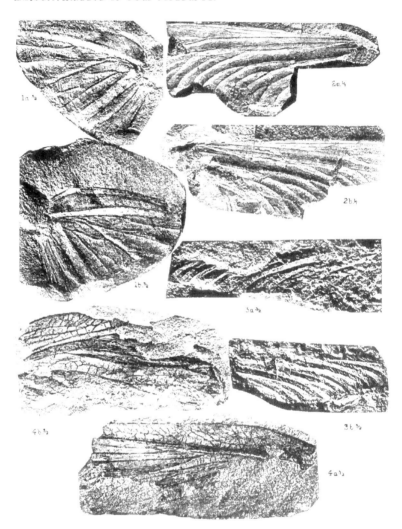

1. DICTYONEURA. 2. ORTHOCOSTA. 3. PTERONIDIA

4. HYPERMEGETHES.

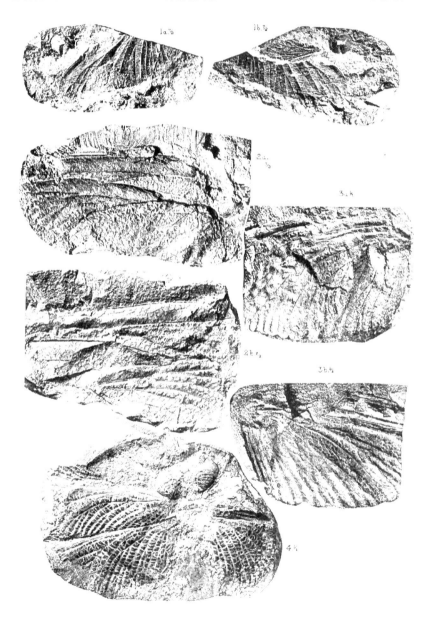

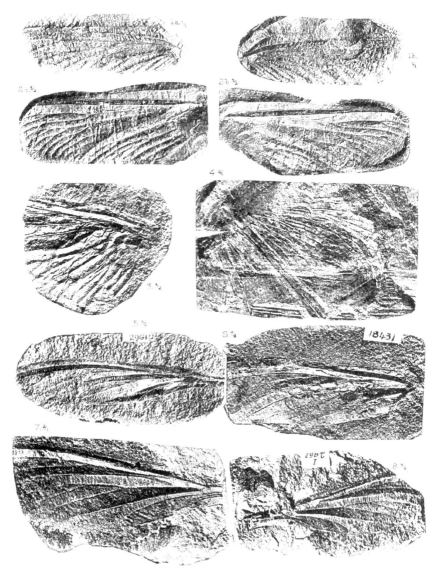

1 LITHOSIALIS. 2 PRUVOSTIA. 3.SPILAPTERA.
4. BOLTONIELLA. 5 8.BRODIA.

Ingram Content Group UK Ltd.
Milton Keynes UK
UKHW020757190723
425424UK00009B/180